This is a first-hand account of one of the most creative and exciting periods of discovery in the history of physics. From 1965 until 1990 theoreticians and experimentalists worked together to probe deeper and deeper into the basic structure of reality, moving closer and closer to an understanding of the ultimate building blocks from which everything in the universe is made.

Gerard 't Hooft worked in the field throughout this period of almost unprecedented discovery, and was closely involved in many of the advances in the development of the subject. In this book he gives a personal account of the process by which physicists came to understand the structure of matter through the development of what is now known as the Standard Theory. In the latter part of the book, he speculates on structures even smaller than those already known to exist, on black holes, grand unification and on possible directions in which the subject may evolve in the future.

This fascinating personal account of the last twenty-five years in one of the most dramatic areas in twentieth century physics will be of interest to professional physicists and physics students, as well as the educated general reader with an interest in one of the most exciting scientific detective stories ever.

D0048248

In search of the ultimate building blocks

In search of the
ultimate building blocks

Gerard 't Hooft

Institute for Theoretical Physics
University of Utrecht, The Netherlands

CAMBRIDGE
UNIVERSITY PRESS

PUBLISHED BY THE PRESS SYNDICATE OF THE UNIVERSITY OF CAMBRIDGE
The Pitt Building, Trumpington Street, Cambridge CB2 1RP, United Kingdom

CAMBRIDGE UNIVERSITY PRESS
The Edinburgh Building, Cambridge CB2 2RU, United Kingdom
40 West 20th Street, New York, NY 10011–4211, USA
10 Stamford Road, Oakleigh, Melbourne 3166, Australia

First published 1997

Printed in the United Kingdom at the University Press, Cambridge

Typeset in Times 9/13pt

A catalogue record of this book is available from the British Library

Library of Congress Cataloguing in Publication data

Hooft, G. 't.
In search of the ultimate building blocks / Gerard 't Hooft.
     p.   cm.
ISBN 0 521 55083 1 (hc). – ISBN 0 521 57883 3 (pbk).
1. Standard model (Nuclear physics) I. Title.
QC794.6.S75H66   1996
539.7′54–dc20   96-31468   CIP

ISBN 0 521 55083 1 hardback
ISBN 0 521 57883 3 paperback

To my mother

To the memory of my father

To Betteke, Saskia and Ellen

Contents

Foreword: an apology

It is difficult to venture into the world of the ultimately small, or even to talk about it, without a very good understanding of the laws of Nature that govern that world. The forces one finds there determine the way in which the tiny particles we wish to study move about, and also all their other properties. Whether, and how, we can actually observe them will also depend on these forces.

And this is not easy, because the laws of Nature are complicated. More and more experts in this field seek refuge in a kind of mathematical gibberish no 'normal' person can follow, unless he or she belongs to the in-crowd. To really appreciate the rock-solid logic of the laws of physics, one actually cannot avoid math. Nonetheless, we physicists feel the need to share the joy of all our beautiful discoveries with whoever wants to listen. We are advised then to avoid all math. This then is what I shall reluctantly do.

It is my intention to provide a narrative of twenty-five years of research on the very tiniest particles of matter. During those twenty-five years, I began to view Nature as an intelligence test to which humanity as a whole has been subjected, as a giant jigsaw puzzle given to us to play with. Time and again, we stumble upon new pieces, large or small, that fit beautifully with those we already have. I want to share with you the sense of triumph we feel at such moments.

So what I need to do is to translate math into plain English. This is certainly possible, but part of the picture will be lost, in particular when I try to formulate 'arguments' commonly used to justify some theory or description, and to reject others. If you, dear reader, feel that you cannot possibly follow my arguments, you may of course hold me to blame, but I will plead in extenuation that translating mathematical formulae into plain language is sometimes impossible without a little cheating.

In many cases, I will not even try to provide a precise explanation. A reader not at home with the subject of theoretical physics may have to accept many of my statements on faith. The idea then is that you get some rough idea about the situation, while avoiding all details of an (often quite extensive) history preceding this wisdom.

The first four chapters form a summary of the accepted understanding of the world of molecules, atoms and atomic nuclei. About the extremely exciting research that led to this knowledge, this book will be brief. In that time its author was not even wearing a diaper, and he prefers to leave an account of that to historians of science.† This book is about what happened next.

In the last twenty-five years, our understanding of the 'elementary particles' has widened so much that some investigators have begun to speculate upon 'the end' of this research: the ultimate theory of all particles and forces, the so-called Theory of Everything, or TOE. Does not even the thought of such an all-embracing theory betray a boundless over-estimation of our capacities, or an equally boundless under-estimation of the infinite complexity of our universe? I will have some things to say about this later. Briefly, the idea may be not as absurd as it sounds.

This book is not meant to be an historical overview of all we know about elementary particles. Such books exist already, and many are very good.‡ Neither has it been my intention to write a systematic introduction to particle physics,§ even though the end product bears some resemblance to that. I have even added a glossary at the end of the book, in response to the request of several readers. This book is intended to be no more than a personal account of some developments in this field. My intention is to share with you my own enthusiasm, and that of my fellow scientists, using terms that I hope will be reasonably understandable to everyone.

All those people mentioned by name are those who, in some way or other, played an important part in developing the picture of the smallest particles of matter in the way that I want to sketch it. They all are people I admire for their contributions, and even in this small selection there will be many omissions. For inaccuracies in mentioning, or not mentioning, names I apologize in advance.

For one bit of national pride, I will probably be pardonned. There are quite a few Dutch investigators who have won their spurs in the subject, and I will

---

† See, for instance, A. Pais, *Inward Bound, Of Matter and Forces in the Physical World* (Oxford, Clarendon Press/New York, Oxford University Press, 1986).

‡ See, for instance, Robert P. Crease and Charles C. Mann, *The Second Creation: Makers of the Revolution in Twentieth-Century Physics* (New York, Macmillan, 1986).

§ See, for instance, Anthony Hey and Patrick Walters, *The Quantum Universe* (Cambridge University Press, 1987), or Leon M. Lederman and David N. Schramm, *From Quarks to the Cosmos, Tools of Discovery* (New York, The Scientific American Library, 1989).

mention many of them. Less forgivable, no doubt, is that my own role will seem bigger than it actually was. This I cannot entirely avoid. After all, I plan to reveal the concept of elementary particles as seen through my own eyes, and of course what is seen up close will appear magnified.

The position I have reached in the world of Physics that has enabled me to give this detailed account would have been unattainable without the strong influence of many people. First, there were my high school teacher, Dr W. P. J. Lignac, and my uncle, Prof. Dr N. G. van Kampen; then my thesis advisor, Prof. Dr M. J. G. Veltman. Each of them shared with me their ideas on the nature of our physical world, and out of their views grew mine. There were also numerous fine physicists with whom I enjoyed many discussions and who helped me feel the intense pleasure of new insights breaking through.

During the completion of this book, I was supported by my family throughout, even if this meant that during vacations much of my attention went to a little notebook computer rather than to them. When I thought I had finished my English translation, I benefitted enormously from the help of Mrs Robin Mize, who assisted me in straightening out my defective English.

1 The beginning of the journey to the small: cutting paper

Let us begin our journey to the world of the tiny by beginning with what we can see with the naked eye, and with those laws of physics that we are all accustomed to. Take a large piece of paper and fold it into an airplane. You may decide to cut the paper in half to make two smaller airplanes. You could even cut the smaller pieces again to make even smaller planes. The properties of the paper, and the rules for folding it into an airplane, will hardly be different, except that the airplanes will become smaller and smaller. Gradually, however, as you continue to cut the paper into smaller and smaller pieces, you will find it harder to make planes, and eventually you will only have little shreds of what once was usable pieces of paper. The property 'foldable into an airplane' has been lost.

We have a similar situation when we start off with a bucket of water and pour the water into smaller buckets. The physical properties of water such as 'flows from high to low' remain the same, until finally we have less than one drop left. You cannot pour drops from high to low; you have to shake them off.

Every child who plays with toy cars or dolls knows that one can imitate the world of the large at a smaller scale. The writer Jonathan Swift based his famous stories on this. An adventurer named Gulliver wanders into the land of Lilliput, which is inhabited by very tiny people. Everything there is tiny: Nature, plants and animals are all scaled down. He himself is seen as a giant there, 'Man-Mountain'. He even manages to extinguish a dangerous fire in the Royal Palace by urinating on it.

During another journey, miraculous forces of fate bring Gulliver to a country named Brobdingnag, where people and all other living and lifeless beings are much larger than Gulliver is used to. There, Gulliver is a dwarf, cherished by a little girl named Glumdalclitch. Finally, Gulliver, in his cage, is picked up by an

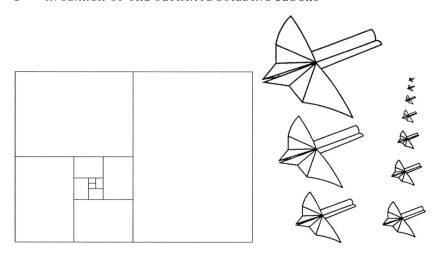

Figure 1. Cut paper, and the airplanes made from it.

eagle which drops him into the sea, and there he is picked up by ordinary-sized sailors, who listen to his story with disbelief.

And they were right not to believe him. No matter how well told, these tales raise intriguing questions. We know, for instance, that little candles have flames nearly as big as large candles. How large were the flames of the candles in Lilliput? And the more you think, the more questions you come up with: how large were the raindrops in Lilliput and Brobdingnag? Were the physical laws for water there different from the ones in our own world? And finally the physicists ask: how large were the atoms in those places? What kind of chemical reactions would they undergo with the atoms of Gulliver's body?

With these questions, the stories falter. The real reason why the worlds of *Gulliver's Travels* cannot exist is that the laws of Nature do not stay exactly the same when you change the scale. This is sometimes evident in disaster movies, where perhaps a tidal wave or a burning skyscraper is mimicked in a scale model. Now, the best results are usually obtained when the scale factor for time is chosen to be the square root of that for size, so, if the skyscraper is built on a scale of 1:9, one has to show the movie at one-third of the actual speed. But even then the skilled eye can make out the differences between what happens in the film and what is observed in the real world.

To summarize, there exist two important aspects of the laws governing our physical world: many laws of Nature stay as they are when you change the scale, but there are phenomena that do not change accordingly, such as a burning candle and drops of water. The ultimate implication of this is that the world of very tiny objects will be entirely different from the ordinary world.

2 To molecules and atoms

Certainly in the world of living creatures scale does create important differences. In many respects, a mouse's anatomy is a carbon copy of that of an elephant, but, whereas a mouse can climb up a nearly vertical piece of rock without much difficulty (and even if it fell from a height many times its own size, it would not greatly injure itself), an elephant would not be able to perform such a feat. Quite generally, the effects of gravity are less important as we study smaller and smaller objects (be they living or inanimate).

Arriving at unicellular creatures, we see that for them there is no distinction at all between up and down. For them, the surface tension of water is a far more important force than gravity. For example, just observe that surface tension is the force that gives a drop of water its shape. Compared with the size of unicellular creatures, drops of water are very big; evidently, surface tension is very important at this scale.

Surface tension is a consequence of the fact that all molecules and atoms attract each other with a force that we call the Van der Waals force. This Van der Waals force has only a very short range. To be precise, the strength of this force over a distance r is roughly proportional to $1/r^7$. This means that if you reduce the distance between two atoms by one-half, the Van der Waals force with which they attract each other becomes $2 \times 2 \times 2 \times 2 \times 2 \times 2 \times 2 = 128$ times stronger. When atoms and molecules come very close to each other, they can bind together very firmly via this force.

Johannes Diderik van der Waals (1837–1923) graduated in 1873 in Leyden with a thesis that would make him famous. It was in Dutch, and was entitled 'Over de continuïteit van de gas- en vloeistoftoestand' (On the continuity of the gaseous and the liquid state). At that time, the existence of molecules and atoms was not at all widely accepted, but Van der Waals showed that the properties of gases and liquids could be very well understood by assuming that these very tiny particles each occupy a certain volume of space, and that as soon as they are separated far enough from each other they all attract one other. The famous

English physicist James Clerk Maxwell, much impressed with this piece of work, remarked that it had prompted quite a few researchers to take up the study of the 'Low Dutch Language'.† In 1910, Van der Waals received the Nobel Prize, but the 'Low Dutch Language' never made it as an internationally accepted language of science, which in previous centuries had been Latin and Greek; later German, French and English. Today, to the regret of some, all science happens in English.

The sizes of unicellular plants and animals are measured in micrometers or 'microns', where 1 micron is 1/1000 of a millimeter, roughly the size of the smallest details one can see through an ordinary microscope. The world of microbes is fascinating, but it is not the subject of this book. We must continue our journey into the world of the small until we arrive at atoms and molecules themselves. At this point, the Van der Waals force must give way to a much more sophisticated kingdom of forces: that of chemistry.

The chemist views atoms as if they are more or less spherical objects, having a diameter of one or several angstroms, where 1 angstrom is $1/10\,000$ of a micron, or 10^{-10} meter (one-ten-billionth of a meter). Practically all the mass‡ of an atom resides in a tiny grain at its center called the nucleus. More about this later.

At small distances, the forces between atoms become extremely complicated; it looks a bit as if they are equipped with hooks and eyes with which they can hold on to each other. The sturdy grouplets of several atoms that can be formed in this way are called molecules.

Consider, for example, the oxygen atom, O. It has two hooks. The hydrogen atom, H, has one eye. One O and two H atoms can be attached to form one molecule of water, H_2O. Two eyes can also grab each other (for example H_2 = hydrogen gas molecule), and so can two hooks (O_2 = oxygen gas molecule), but this coupling is not quite as firm.

The two hooks on the oxygen atom are not situated quite opposite to each other, but at an angle of about 104 degrees. A similar thing can be said for many of the other atomic species. Consequently, the molecules acquire a complicated shape. One of the nicest building blocks is the carbon atom, which has four eyes that attach well to the eyes on other carbon atoms. Many of the molecules in living beings are built from chains of carbon atoms, C.

There are over one hundred different species of atoms, and each of these

---

† See H. B. G. Casimir, *Haphazard Reality* (New York, Harper & Row, 1983).

‡ A reader not yet familiar with the notion of 'mass' may use as a rule of thumb the concept that all objects on Earth have masses equal to their weights (masses are also measured in grams or kilograms), but that weight is the force with which the Earth pulls an object towards the ground. In a spaceship your mass is the same as that on Earth, but your weight is practically zero.

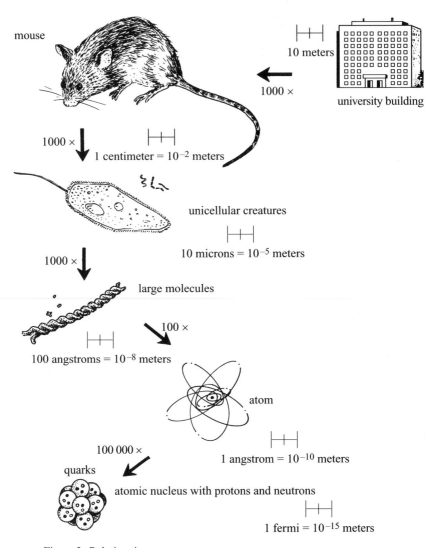

Figure 2. Relative sizes.

species exerts forces that are characteristic for that atom and differ to a larger or lesser degree from what the other atoms do. Substances consisting of only one species of atoms are called the chemical elements. The word 'atom' derives from the Greek ἄτομος, meaning 'undividable', and the use of the word 'element' suggests that we have arrived at the fundamental building blocks of matter.

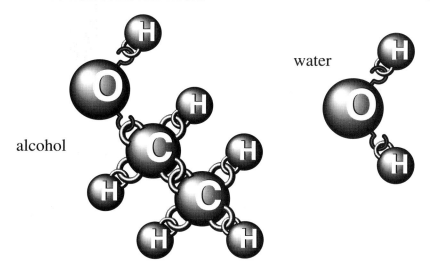

Figure 3. Atoms can be connected together as if they have hooks and eyes.

Indeed, this was how it looked in the middle of the nineteenth century when these terms were invented. This was a mistake, as we know now, because atoms can be split, and so the elements are no longer truly elementary. We hang on to this nomenclature, even if it is formally incorrect, because we are now so accustomed to it. But do not expect that humanity has learned its lesson! The words 'elementary particles' are just as clumsy. Similarly, what should one think about the terms 'modern music' and 'post-modern music', etc? Surely a time will come when we will regret the introduction of such terminology.

Perhaps you will find my picture of atoms as little spheres with hooks and eyes not sufficiently scientific. True, we speak of the 'chemical binding force', when we refer to the ways atoms can bind to each other. My description of these forces was somewhat whimsical because there is a remarkable aspect to them: the laws of Nature responsible for the chemical binding forces are completely known! Now this statement may well come as a surprise to you. So, is all of chemistry finished and done with, you may ask, even though the newspapers have never mentioned anything of such a revolutionary discovery? Well, no, is my reply, it is merely the fundamental equations underlying the chemical binding forces that are completely known, but unfortunately all calculations starting from these equations are so tremendously complicated that we are forced to use approximation techniques. The accuracy of these mathematical techniques is not always easy to judge, and is often very poor. Even the simplest molecules, such as those of water or alcohol, can often better be studied by doing simple experiments with the substances themselves than by doing *ab initio* calculations starting with our equations. These days, mathematical approximation techniques are very advanced, and what they tell us is that a description of atoms as if they were spheres with hooks and eyes may not be so bad after all.

But of course hooks and eyes are not really there. What gives an atom its nearly spherical shape is the electrons, the electrically charged particles frolicking around the nucleus. An electron is very light: its mass is only about $1/1836$ of that of the lightest nucleus (that of hydrogen). The electric charge of an electron is opposite to that of the nucleus, so although electrons are strongly attracted to a nucleus, they repel each other. If the total electric charge of the electrons in an atom matches that of the nucleus, the atom is said to be in equilibrium. Usually, several electrons are needed for this. We then say that the atom is electrically neutral.

The force controlling electrons is actually mathematically quite simple; we call it the electrostatic, or Coulomb, force. Yet it is the electrons to which an atom owes the remarkable properties that we referred to as 'chemical binding'. This is because the laws of motion for the electrons are so very special. Their movements are governed entirely by 'quantum mechanics'. The theory that we call quantum mechanics was completed in the early twentieth century. It is a paradoxical theory, difficult to understand or explain, but exciting, fantastic and revolutionary. Present-day theoretical physics is difficult to imagine without quantum mechanics, and quantum mechanics is very much at the center of all of theoretical elementary particle physics. I shall have much more to say about quantum mechanics later, but I will not make any attempt to explain the chemical binding forces using this theory (though it is possible!)

Not only electrons, but also atomic nuclei, and atoms as a whole, obey the laws of quantum mechanics, but, because nuclei and atoms are much heavier than electrons, the consequences of this are much less drastic. For all intents and purposes, the chemist may treat atoms as if they were ordinary billiard balls, with very special forces between them only arising when they approach each other very closely.

3 The magical mystery of the quanta

Twentieth century physics began exactly in the year 1900 when the German physicist Max Planck proposed a possible solution to a problem that had been haunting physicists for years. The problem concerned the light emitted by any object that is heated to a certain temperature, and also the softer infrared radiation emitted, with less intensity, by colder objects.

It had become well accepted that all this radiation had an electromagnetic origin, and the laws of Nature for these electromagnetic waves were already known. Also known were the laws for heat and cold, so-called 'thermodynamics'. Or so it seemed. But if we use thermodynamic laws to compute the intensity of the radiation in question, the outcome does not make any sense at all. Such a calculation would tell us that an infinite amount of radiation should be emitted in the very far ultraviolet. Of course, this is not what happens. What does happen is that the intensity of the radiation shows a peak at a certain characteristic wavelength, and diminishes both at wavelengths longer and shorter than this. This characteristic wavelength is inversely proportional to the absolute temperature of the radiating object (absolute temperature is defined by a temperature scale that begins at 273 °C below zero or 459.6 °F below zero). At 1000 °C or 1800 °F an object is 'red hot'; at this point the object is radiating in the visible light region.

What Planck proposed was simply that radiation can only be emitted in packages of given sizes. The amount of energy in one such a package, or quantum, is inversely proportional to the wavelength, and hence proportional to the frequency of the emitted radiation. The formula is:

$$E = h \times v,$$

in which E stands for the energy in one package, v is the frequency, and h is a new fundamental constant of Nature, Planck's constant. When Planck computed

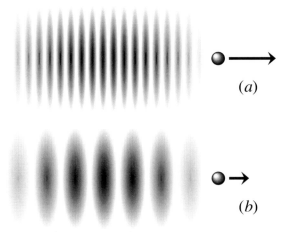

Figure 4. Particles are associated with waves. Energetic particles (*a*) have shorter and more rapidly oscillating waves than less energetic ones (*b*).

the intensity of heat radiation imposing this new constraint, the result agreed perfectly with observations.

Shortly after this, in 1905, Einstein formulated this theory in a much more radical way: he suggested that not only do heated objects *emit* radiation in energy packages, but that radiation must entirely *consist* of multiples of Planck's energy packages. Following this, the Frenchman Prince Louis-Victor de Broglie† turned this around once again: he theorized that not only does anything that oscillates exist as given amounts of energy, but anything with energy must behave like a 'wave' spread over some region in space, which oscillates with a frequency v as dictated by Planck's equation. The quantum that is associated with light rays would henceforth be viewed as one of the various species of elementary particles: the photon. All other species of particles have different kinds of oscillating waves of force fields associated to them, but more about this later.

The curious behavior of electrons inside an atom, which was discovered and elaborated upon by the famous Danish physicist Niels Bohr, could be attributed to de Broglie's waves. After this, it was Erwin Schrödinger who discovered, in 1926, how to transform de Broglie's wave theory into accurate mathematical equations. The precision with which all the calculations could be performed was

† Pronounced as: *Debroiye.*

startling, and soon it became clear that the behavior of *all* small objects was determined accurately by the newly discovered 'quantum wave equations'.

Quantum mechanics works beautifully, there is little doubt about that. However, a very peculiar question presented itself: *what* do these equations actually mean? What is it that they are describing? When Isaac Newton, back in 1687, formulated how planets should move around the Sun, it was clear to everyone what his equations were saying. Planets are always at some well-defined position in space, and their positions and their velocities at one particular moment determine unambiguously how positions and velocities will vary as time goes by.

But for electrons everything looks very different. Their behavior seems to be shrouded in mystery. It seems as if electrons can 'exist' at different places simultaneously. They are cloud-like, wave-like, and this is not just a small effect. If one carries out sufficiently accurate experiments, one can determine that the electron seems to be able to move simultaneously over trajectories that are very far apart from each other. What can all this possibly *mean*?

Niels Bohr managed to answer this question in a way one can work with, and many physicists, to this day, find his answer satisfactory. It is called the 'Copenhagen interpretation' of quantum mechanics. Instead of saying that the electron sits at a point **x**, or at a point **y**, we talk about the *state* of an electron. Then we not only have the state '**x**', or a state '**y**', but also states such as 'partly **x** *and* partly **y**'. A single electron can therefore be at many places simultaneously. Quantum mechanics tells us very precisely how such an electron state changes as time goes on.

A 'detector' is a piece of apparatus with which one can determine whether or not a particle is present somewhere. It could be a particle counter, or a sensitive piece of celluloid, or even the human eye. If a particle meets the detector, the state of the particle will be disturbed, so only if we do not want to study how the state evolves further are we allowed to put a detector in its way. If we know what the quantum state is, we can compute the *probability* that the detector will register the particle at a point **x**. If the particle is indeed registered there, then from that moment† the particle sits in the state '**x**'.

The laws of quantum mechanics have been formulated very accurately. We

† That the quantum state of a particle 'jumps' just because it is being detected is sometimes presented as if it were a special axiom of quantum mechanics. Actually this jumping is nothing but a severely simplified description of the complicated interaction processes that take place in a detector. 'Detection' really means that, depending on a microscopic phenomenon (such as the movement of a single particle), a very large number of molecules in the 'detector' (like a needle or an indicator light) are moved into different states.

know exactly how to compute anything we would like to know. But if we wish to 'interpret' our results, we have to deal with a curious, fundamental *uncertainty*: that various properties of the tiny particles cannot be simultaneously well defined. For instance, we can fix the velocity of a particle very precisely, but then we will not know any more exactly where it is, or, alternatively, we can determine its position accurately, but then its velocity is ill-defined. If a particle has 'spin' (rotation about its axis) then the direction in which it is spinning (the orientation of its axis) cannot be defined with great precision.

It is not easy to explain in simple terms where this uncertainty comes from, but there are examples in ordinary life where we have something similar. The *pitch* of a tone and the *moment in time* at which you hear the tone have similar mutually connected uncertainties. If one wishes to tune a musical instrument, one must listen to a note for some time, and compare it, for instance, with a tuning fork, which must also be kept vibrating for some time. Very short notes do not have a precisely defined pitch, so, for example, if one plays very short staccato notes, one cannot hear very well whether the instrument is well tuned. This holds in particular for the lowest notes. I am talking here about a fundamental property of sound having nothing to do with musicality. You might object that a skilled musician can tell how well the instrument is tuned even if you play short notes, but that is then because he knows the instrument so well that he can judge this by the overtones, their pitches being better defined.

The rules of quantum mechanics work, but only if *all* natural phenomena in the world of the small are subjected to the *same* rules. This includes viruses, bacteria, even people. However, the bigger and heavier an object is, the harder it becomes to observe the quantum mechanical deviations from the ordinary, 'classical' laws of movement. I would like to refer to this peculiar and important demand in the theory as 'holism'. This is not quite the same thing as what some philosophers mean when they use the word 'holism', which would be 'The whole is more than the sum of its parts'. Well, if physics has taught us one thing, it is quite the opposite of that: an object composed of a large number of particles can be *exactly* understood if you know the properties of its parts (the particles), if only one does the sum correctly (and this is far from easy in quantum mechanics!) What *I* mean by holism is that, yes, the whole is the sum of its parts, but you can do the sum only if all parts obey the same laws. For instance, Planck's constant, $h = 6.626075\ldots \times 10^{-34}$ joule seconds, must be exactly the same for everything everywhere. It must be a *universal* constant.

The rules of quantum mechanics work so well that it has become very hard to refute them. Ingenious tricks discovered by Werner Heisenberg, Paul Dirac and many others further improved and streamlined the general rules. But Einstein, as well as other early pioneers such as Erwin Schrödinger, always had serious objections to this interpretation. Maybe it works all right, but where exactly *is* the electron, at the point **x** or at the point **y**? In short, where is it *in reality*? What is the reality that is hidden behind our formulae? If we are to believe Bohr, it is senseless to search for such a reality. The quantum mechanical rules by themselves, and the actual observations performed by the detectors, are the only realities we are allowed to talk about.

To this day, many researchers agree with Bohr's pragmatic attitude. The history books say that Bohr has proved Einstein wrong. But others, including myself, suspect that, in the long run, the Einsteinian view might return: that there is something missing in the Copenhagen interpretation. Einstein's original objections could be overturned, but problems still arise if one tries to formulate the quantum mechanics of the entire universe (where measurements can never be repeated), *and* if one tries to reconcile the laws of quantum mechanics with those of gravitation. But I am running far ahead in my story (I will return to this point in chapter 28). For a correct description of atoms and molecules, quantum mechanics is a perfect theory.

The elusive mystery of quantum mechanics gave rise to a great deal of controversy, and the amount of nonsense that has been claimed is so voluminous that a sober physicist does not even know where to start to refute it all. Some claim that 'life on Earth started with a quantum jump', that 'free will' and 'consciousness' are due to quantum mechanics. Even paranormal phenomena have been ascribed to quantum mechanical effects.

I strongly suspect that all this is an attempt to ascribe 'unintelligible' phenomena (such as the paranormal) to 'unintelligible' causes (such as quantum mechanics). But quantum mechanics is not unintelligible at all, and the theory itself provides the counter-arguments. Its 'holistic' character, in the sense I explained earlier, implies that the outcome of any calculation is *always* a *chance* or *probability*. If an experiment is repeated many times, a certain percentage of the results will be one outcome, and another percentage a different outcome. The more often an experiment is repeated, the more accurately these percentages will agree with calculations using the quantum mechanical laws. So this should also hold for the 'emergence of life', a 'decision taken out of free will' or some 'paranormal experience', if it had been possible to repeat such 'experiments' many times. Of course, no human being can possibly 'compute' such phenomena from their elementary particle building blocks, but, if one could, these phenomena would have to be considered just as 'experimental observations'.

One should not underestimate how forbidding the large numbers of particles are that prohibit any attempt to calculate large scale phenomena exactly from their quantum mechanical constituents, but here we can argue that no brain cell or chemical reaction exists that can bypass the quantum mechanical uncertainty relations. If a brain cell tried to make a 'paranormal' calculation, it would necessarily make many mistakes. The outcome of its calculations should be fully in accord with the quantum uncertainty relations. And this also holds for the origin of life. Perhaps life on Earth originated as a result of an extremely improbable coincidence of events, but this has nothing to do with quantum mechanics. Very many people seem to cherish some deeply felt desire for the unknown, for mysticism, and quantum mechanics seems to fulfil this desire. Not for me. Quantum mechanics is a logically coherent theory for the atomic forces and movements. Physicists should see it as their task to combat obscurantism, but some of us are perhaps not so aware of this. Niels Bohr, for instance, used the well-known Yin–Yang symbol to symbolize complementarity in quantum mechanics; that is, the fact that a particle must sometimes be treated as a wavelet and sometimes as a real particle. He did not mean to say that meditation and contemplating one's navel would help us towards deeper insight in the mysteries of quantum mechanics, as some want us to believe.

The real nature of quantum mechanics, I think, may be summarized as follows. In principle, the outcome of any experiment can be foretold by using our present understanding of the natural laws, in the sense that the prediction will consist of two factors. The first factor is some precisely defined calculation of the effects of forces and structures, which is as rigorous as Isaac Newton's laws for the movements of planets in the solar system. The second factor is a strictly mathematically defined, uncontrollable, statistical arbitrariness. Particles will follow a given probability distribution first this way, then that. The probabilities are computable, and so are the chances that a single experiment will give *deviations* from the computed probabilities, and so on.

Probabilities and statistics are mistreated a great deal, even by physicists. Some have uttered, for instance, the theory that all possibilities for certain events are being realized in 'parallel worlds', with their given probabilities. This is called the 'many worlds' interpretation of quantum mechanics. This is how crazy it becomes if one tries to 'quantize the universe'. To my sober mind, all this is nonsense. Much more reasonable is the suspicion that the statistical element in our predictions will eventually disappear completely as soon as we know exactly *the complete theory of all forces*, the Theory of Everything. This implies that our present description involves variable features and forces which we do not (yet) know or understand. Such an interpretation is called the 'hidden variables hypothesis'.

Numerous attempts have been made to elaborate on this idea in terms of certain mathematical models. When this continued to be unsuccessful, physicists did what they always do under such circumstances: they proved that it is impossible. Albert Einstein, Nathan Rosen and Boris Podolsky constructed a 'Gedankenexperiment', a hypothetical experiment only performed on paper, for which quantum mechanics predicted an outcome that they claimed to be impossible to reproduce in any reasonable hidden variable theory. Later, the Irish physicist John Stewart Bell succeeded in elevating this result into a mathematical theorem. If you start off with switches and gears, or whatever, you can never construct a universe in which you see quantum mechanical phenomena, according to Bell. We call such a thing a 'no-go theorem'.

You may have already suspected that I still believe in the hidden variables hypothesis. Surely our world must be constructed in such an ingenuous way that some of the assumptions that Einstein, Bell and others found quite natural will turn out to be wrong. But how this will come about, I do not know. Anyway, for me, the hidden variables hypothesis is still the best way to ease my conscience about quantum mechanics. And as for 'no-go theorems', we will encounter several of these and discuss their fate.

4 Dazzling velocities

Nearly 100 000 times smaller than an atom itself is the small grain at its center: the atomic nucleus. By its mass, but even more by its electrical charge, the nucleus determines all properties of the atom of which it makes part. Because of the sturdiness of the nucleus, it seems as if the atoms that give shape to our daily world are themselves unchangeable, even if they interact with each other to form chemical substances. But sturdy as it is, a nucleus *can* be torn apart. If atoms are smashed against each other with dazzling velocities, it may happen that two nuclei hit each other, and then they can either break into pieces or coalesce. At the same time, *subnuclear particles* are released. The new physics of the first half of the twentieth century was dominated by the new riddles these particles presented.

But we have quantum mechanics, you will say, and is that not applicable everywhere? What's the difficulty? Indeed, quantum mechanics does apply to the subatomic particles, but there is more to it than that. The forces these particles exert on one other, and which keep the atomic nucleus together, are so strong that the velocities with which they swirl around each other, inside and outside the nuclei, come close to the speed of light, which is about 300 000 kilometers/second, or 186 000 miles/second. When such velocities come into play, a second modification is needed to the nineteenth century laws of physics: we have to take account of Einstein's *special theory of relativity*.

This theory also was a result of a publication by Einstein in 1905. His starting point was that an experiment carried out in a laboratory in outer space should yield results that do not depend on how fast and in which direction the laboratory is moving, even if one tried to measure the speed of light in that laboratory. This is odd. Suppose that a spaceship has a velocity of 50 000 kilometers/second. You might expect that in one direction the light velocity will amount to 350 000 kilometers/second and in the opposite direction have a reduced value of 250 000 kilometers/second. You would also expect a small deviation from the normal 300 000 kilometers/second in the orthogonal direction.

The point is that performing such an experiment depends on the existence of accurate clocks and yardsticks. Furthermore, various clocks must have been synchronized. That clocks and yardsticks may very well be affected by the velocity of the laboratory had already been suggested by the Dutchman Hendrik Antoon Lorentz (1853–1928), and, independently of him but a few years earlier (in 1889), by the Irishman George Francis Fitzgerald. To many people in the Netherlands, Lorentz is better known for other reasons: he presided over a committee whose task it was to assess the possibilities of closing off a large body of water in the Netherlands, the Zuyderzee. A 20 mile barrier dam had to be designed to separate the Zuyderzee from the North Sea. Water currents due to tidal movements had to be calculated. Considering the fact that in those times no computers existed, Lorentz's calculations would later turn out to be remarkably accurate.

As for static and moving clocks and yardsticks, Lorentz had regarded them as being *affected* by their motion. It was Einstein who realized fully that with these effects all movement and rest have become *relative* notions. There is no such thing as absolute rest, or an absolute reference frame with respect to which one can measure the speed of light.

There was more that had to become relative. *Mass*† and *energy* in this theory also depend on velocity, as do electric and magnetic field strengths. Einstein discovered that the mass of a particle is always proportional to the energy it contains, provided that you take into account a very large amount of 'rest energy' for each particle. A particle's rest energy is proportional to its mass if it is at rest:

$$E = M \times c^2.$$

Here, E is the particle's energy, M is its mass and c is the velocity of light, a universal constant.

This equation suggests that every particle must harbor enormous quantities of energy, since the speed of light is very great, and it was partly this prediction that would make relativity theory very important for physics (and indeed for everyone!) Now, the theory of relativity will also be self-consistent only if it is 'holistic', that is if everything and everybody obeys the laws of relativity. Not only do clocks tick at slower rates when they move at great speeds, but all processes, lifeless as well as animate, behave in the unusual way described by this theory

---

† That is, if mass M is defined by Newton's law $F = M \times a$. Modern physics teachers prefer to redefine mass such that it is velocity independent.

(a)

(b)

Figure 5. According to Einstein's special theory of relativity, the way we perceive time and distance depends on where we are and how we move. Someone inside a fast moving train (a) may have synchronized his clocks and measured the length of his train, but someone outside (b) may find that these clocks are not showing equal time, and that the train is somewhat shorter, though for him the trees are farther apart.

when approaching the speed of light. The human heart is simply a biological clock, and it will beat at a slower rate when travelling in a near-light-speed spaceship than it will on Earth. This strange phenomenon leads to what is known as the 'twin paradox' suggested by Einstein, in which identical twins age differently if one remains on Earth whilst the other travels at speeds approaching that of light.† Neither twin, however, would be able to determine the absolute velocity of the laboratory he or she was in.

† The one who feels the acceleration of the engine of his spacecraft will end up being the younger of the two. The fact that the other one may feel the Earth's *gravitational field* can only be accommodated for in the *general* theory of relativity.

The electromagnetic force

The *electric force* is the force with which two charged particles repel each other (if their charges are equal), or attract each other (if their charges are of opposite sign).

The *magnetic force* is the force experienced by a *moving* electrically charged particle when it goes through a magnetic field. A magnetic field arises when electrically charged particles move, for instance if electrons flow through the windings of a coil.

Electric and magnetic forces are interlinked. In 1873, James Clerk Maxwell succeeded in formulating the complete equations that electric and magnetic force fields obey. In this way, he obtained a 'unified theory', now called electromagnetism.

Characteristic properties of the electromagnetic force, as it acts on elementary particles, are the following.

- The force acts in a universal way on something that we call the *electric charge*.
- The force has a very large range (magnetic fields can extend between stars).
- The force is *fairly weak*. Relevant for its strength is the ratio between the square of the charge of an electron and $2hc$ (twice Planck's constant times the speed of light). This ratio is found to be approximately 1 : 137.036.
- The 'carrier' of this force is the photon, a particle with vanishing mass (when at rest) and spin 1.† The photon itself has no electric charge.

Quite generally, the range of a force is inversely proportional to its carrier's mass.

---

† The notion of *spin* is explained in the next chapter.

The gravitational force

Gravity is a very fundamental force for which Einstein discovered the detailed structure in 1915, connecting it to curvature in the fabric of space and time. However, even to this day we do not know how to reconcile the laws of gravity with those of quantum mechanics (except where the gravitational force is sufficiently weak).

- The gravitational force acts exclusively on a particle's *mass*.
- The force has an extremely large range (it can probably act all the way to the farthest realms of the known part of the universe).
- The force is so extremely weak that it will probably never be possible to detect experimentally the mutual gravitational attraction between two elementary particles. The only reason why we can experience this force is because it is *collective*: all particles (in the Earth) pull all particles (in our body) in the same direction.
- The carrier is the hypothetical 'graviton'. Although not yet discovered experimentally, we do know what quantum mechanics will postulate: that it has zero mass and spin 2.

A general rule for forces is that, if the carrier has *even* spin, the force between like charges will be attractive and between opposite charges repulsive. If the spin is odd (as in electromagnetism), the converse is true.

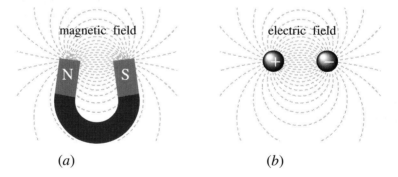

(a)   (b)

Figure 6. Maxwell's equations give rise to magnetic field lines (a) in the same shape as electric field lines (b).

Einstein quickly realized therefore that the laws of gravity would also have to be adjusted to obey the relativity principle. You will remember that, for small objects, gravity plays a minor role. As experienced by subatomic particles, gravity is extraordinarily weak. So, for our story, gravity does not matter very much. (We will, however, encounter this curious and extremely fundamental force later in this book.) But the problem Einstein was confronted with would also turn out to be important for understanding the other forces between the tiny particles. Therefore I will now disclose where he found the solution, after a ten year search.

In order to apply the principle of relativity to the gravitational force, the principle had to be extended as follows. Not only should it be impossible to determine the absolute velocity of your laboratory, but also it should be impossible to distinguish *changes* in this velocity from the effects of gravitational forces.

Einstein realized that the consequence of this was that gravity does to space and time what moisture does to a flat sheet of paper: it makes it bumpy. Curves and wrinkles appear that cannot be ironed away. Now, the mathematics of curved spaces was already known, but in Einstein's time applying such fancy abstract notions of mathematics to formulate laws of physics was something new altogether, and it took him years to become familiar with it. Nowadays, three-quarters of a century later, physicists are quite used to the practice of flirting with advanced math. But, even today, the problem is not only to deal with abstract mathematics; more often the most difficult part is setting up the *right* mathematical equations and formalisms. Once we have the equations, we can disentangle them and solve them using computers, for instance. But what *are* the equations?

Einstein's theory of gravity is called the *general theory of relativity*. We will return to it later, first because the theory serves as an example for other theories for fundamental forces, and secondly because objects *very much smaller* than the subatomic particles become more sensitive again to the gravitational force. And for understanding the ultimate theory of particles and forces, gravity will be essential. But first we shall discuss the subatomic particles themselves. For their understanding, the *special* theory of relativity, which is the relativity theory without the gravitational force, suffices.

5 The elementary particle zoo before 1970

Our journey to the very small has now brought us beyond the atoms, which are bulky and fragile objects compared with what we shall be occupied with next: the atomic nucleus and whatever is inside. The electrons, now seen 'at a great distance' circling around the nucleus, are themselves also small and extemely robust. I now invite you to have a look inside the nucleus, through the eyes of the scientists before 1970. I consider the years around 1970 as a crucial period, but I am now also choosing the year 1970 because this was the time I became acquainted with elementary particle physics myself, as a young graduate student at the State University of Utrecht in the Netherlands.

All the physics that I mentioned earlier (and of course a great deal more) was basic stuff for students of theoretical physics. Also, a lot was known about the structure of the atomic nucleus. The nucleus is built from two species of building blocks: *protons* and *neutrons*. The proton (Greek $\pi\rho\tilde{\omega}\tau\sigma\varsigma$ = first) owes its name to the fact that the simplest atomic nucleus, that of hydrogen, consists of just one proton. It carries one positive unit of charge. The neutron resembles the proton as if it were its twin brother: its mass is practically the same, its spin is the same, but the electric charge is absent in a neutron; it is neutral.

The masses of these particles will be expressed in units called mega-electron-volts, or MeV for short. $1\,\text{MeV}$ ($= 10^6$ electron-volts) is the amount of energy of motion a singly charged particle (such as an electron or a proton) picks up if it traverses an electric potential difference of 10^6 (one million) volts. Since this energy is also transformed into mass, the MeV is a useful unit of mass for an elementary particle.

Most atomic nuclei contain somewhat more neutrons than protons. Now, the protons, crammed so close together in the tiny nucleus, should repel each other fairly strongly due to their similar electrical charges. However, there is a force holding them tightly together which is much stronger than the electromagnetic force: the so-called *strong force*.

Light (see Chapter 3) is an electromagnetic phenomenon, and it is quantized in 'photons'. Photons quite generally behave as the 'messengers' of all electromagnetic interactions. Likewise, the strong force also has its quanta. The properties of the quantum particles associated to the strong force were predicted by the Japanese physicist Hideki Yukawa (1907–1981); these particles would later be called *pions*. There is a very important difference between pions and photons. A pion is a chunk of matter with a certain amount of 'mass'. If this particle is at rest the mass is always the same, being about 140 MeV (see Table 1 on page 24). If it moves very fast, its mass appears to increase.† In contrast to this, the photon is said to have a vanishing rest mass. By this, we do not mean to say that the photon mass is zero, but rather that a photon cannot be at rest. Like all particles with vanishing rest mass, the photon moves exclusively with the speed of light, some 300 000 kilometers/second, a velocity that the pion can never reach, because this would require an infinite amount of kinetic energy. For the photon, all of its mass can be attributed to its kinetic energy.

The pion is lighter than the proton and the neutron, but heavier than the electron (see Table 1). There are three kinds of pions: one is electrically positively charged, one is negative and the third is neutral. If protons and neutrons strike each other with a sufficient amount of energy, pions are often released. This can be compared with what happens if ordinary atoms are shaken about so that they hit each other frequently, such as what happens when you *heat* a substance. The atoms then begin to emit light, that is, photons. At a scale 10 000 times smaller, atomic nuclei can do the same thing. If they are shaken frantically they spill pions.

The discovery of the pion, incidently, did not go smoothly. It took place at a time (1935) when scientists could not yet produce particles of such a type artificially. They were, however, able to study the subatomic fragments that Nature provides free from the far ranges of the universe: 'cosmic rays'. Cosmic rays were known to leave traces in an apparatus called a 'cloud chamber', and by studying these traces the properties of the particles that had left those traces could be determined. And, indeed, something was found with a mass that agreed reasonably well with what Yukawa had predicted. It was named *meson* (Greek μέσος = middle), because its mass was between that of the electron and the proton. 'Meson' was also what Yukawa had called the particle he had predicted. But there was one important discrepancy: the particle that was observed does not participate at all in strong interactions, and therefore could not possibly be

---

† See footnote on page 17.

Table 1. *The elementary particles with lifetimes longer than 10^{-20} seconds, as they were known up until 1970[a].*

Name	Symbol	Mass (MeV)	Charge	Spin	Lifetime (seconds)	Principal decay modes	$S,\ I_3$[b]
Photon	γ	0	0	1	∞	stable	
Leptons ($L=1$, $B=0$):							
Electron	e^-	0.5109990	−	$\tfrac{1}{2}$	∞	stable	
Muon	μ^-	105.6584	−	$\tfrac{1}{2}$	2.1970×10^{-6}	$e + \bar{\nu}_e + \nu_\mu$	
e-neutrino	ν_e	~ 0	0	$\tfrac{1}{2}$	$\sim \infty$	stable	
μ-neutrino	ν_μ	~ 0	0	$\tfrac{1}{2}$	$\sim \infty$	stable	
Mesons ($L=0$, $B=0$):							
Positive pion	π^+	139.570	+	0	2.603×10^{-8}	$\mu^+ + \nu_\mu$	0, 1
Negative pion	π^-	139.570	−	0	2.603×10^{-8}	$\mu^- + \bar{\nu}_\mu$	0, −1
Neutral pion	π^0	134.976	0	0	0.84×10^{-16}	2γ	0, 0
Positive kaon	K^+	493.68	+	0	1.237×10^{-8}	$\mu^+ + \nu_\mu;\ \pi^+ + \pi^0;\ 3\pi$	1, $\tfrac{1}{2}$
Negative kaon	K^-	493.68	−	0	1.237×10^{-8}	$\mu^- + \bar{\nu}_\mu;\ \pi^- + \pi^0;\ 3\pi$	−1, $-\tfrac{1}{2}$
K-long	K_{Long}	497.7	0	0	5.17×10^{-8}	$3\pi^0;\ \pi^+ + \pi^- + \pi^0$	±1, $\pm\tfrac{1}{2}$
K-short	K_{Short}	497.7	0	0	0.893×10^{-10}	$2\pi^0;\ \pi^+ + \pi^-$	±1, $\mp\tfrac{1}{2}$
Èta	η	547.5	0	0	5.5×10^{-19}	$3\pi;\ 2\gamma$	0, 0

Baryons ($L = 0$, $B = 1$):

							S	I_3
Proton	p	938.2723	$+$	$\frac{1}{2}$	∞	stable	0,	$\frac{1}{2}$
Neutron	n	939.5656	0	$\frac{1}{2}$	887	$p + e + \bar{\nu}_e$	0,	$-\frac{1}{2}$
Lambda	Λ	1115.68	0	$\frac{1}{2}$	2.63×10^{-10}	$p + \pi^-, n + \pi^0, p + e^- + \bar{\nu}_e$	-1,	0
Sigma-plus	Σ^+	1189.4	$+$	$\frac{1}{2}$	0.80×10^{-10}	$p + \pi^0; n + \pi^+$	-1,	1
Sigma-zero	Σ^0	1192.5	0	$\frac{1}{2}$	7.4×10^{-20}	$\Lambda + \gamma$	-1,	0
Sigma-minus	Σ^-	1197.4	$-$	$\frac{1}{2}$	1.48×10^{-10}	$n + \pi^-, n + e^- + \bar{\nu}_e$	-1,	-1
Ksi-zero	Ξ^0	1314.9	0	$\frac{1}{2}$	2.9×10^{-10}	$\Lambda + \pi^0$	-2,	$\frac{1}{2}$
Ksi-minus	Ξ^-	1321.3	$-$	$\frac{1}{2}$	1.64×10^{-10}	$\Lambda + \pi^-, \Lambda + e^- + \bar{\nu}_e$	-2,	$-\frac{1}{2}$
Omega-minus	Ω^-	1672.4	$-$	$1\frac{1}{2}$	0.82×10^{-10}	$\Lambda + K^-; \Xi^0 + \pi^-; \Xi^0 + e^- + \bar{\nu}_e$	-3,	0

[a] Associated to each *lepton* and each *baryon* we also have the corresponding *antiparticles*. The antiparticles have the same mass as the corresponding particles, but their electric charges and the quantum numbers B and L (to be explained shortly) are opposite. The neutral *mesons* are their own antiparticle, and π^+ and π^- are each other's antiparticle, just like K^+ and K^-. The symbol for an antiparticle is the same as that of the particle, with a bar on top. (Masses and lifetimes corrected according to the listings of 1994.)

[b] The symbols S ('strangeness') and I_3 ('isospin') are explained at the end of Chapter 5.

the predicted pion. These days we use for this particle only the abbreviation μ or the name *muon*. This is because in all other respects the disagreement had become embarrassing. The word 'meson' is now used only for pions and related particle types.

Nowadays we know what caused this discrepancy, and how the muon got into the cosmic rays in the first place. High up in the atmosphere, atoms from outer space collide, with enormous energies, with atoms resident in the atmosphere. There they immediately produce lots of pions, in complete agreement with Yukawa's theory. But what Yukawa could not have known is that pions are not stable: neutral pions decay into photons. Every charged pion decays in less than one-ten-millionth of a second into a muon and a *neutrino* (another specimen of the particle zoo; see Table 1). The neutrino usually escapes undetected, but the muons reach the lower parts of the atmosphere and can even be detected hundreds of meters under the ground. Simply because muons do not interact strongly, they can pass through the air and some of the soil in one piece.

When all this became clear, it was Isidore I. Rabi† who neatly summarized the reaction of the scientific world to the discovery of the muon by asking: 'Who ordered *that*?' Even with today's understanding of elementary particles, it would not really have been possible to 'order', or predict, the muon. In any case, nobody has come forward with a credible theory that tells us how one should have computed and predicted the mass of this muon (which turns out to be approximately 200 times the electron mass).

And there was more to come that was not predicted. It was not only pions that were discovered high in the atmosphere. The *kaons* are particles heavier than pions but, in other respects, rather similar. Quite a bit of pioneering work (by the American physicist Murray Gell-Mann and the Dutchman Abraham Pais, among others) was needed to figure out how these particles fit in with the rest. Mysteriously, different species of particles appeared that were all electrically neutral with a mass of around 500 MeV. These particles could be produced in a number of ways, and they decay in various different modes. That they finally turned out to be just two different kinds of particles, K_{Long} and K_{Short}, was not at all obvious in the beginning. In the next chapter, I will explain why these particles are so special.

A noteworthy property of all these tiny particles is that they may rotate about

---

† Rabi was one of the discoverers of the magnetic resonance method, used to study many properties of atoms and molecules in a magnetic field. He was also one of the founding fathers of the European subatomic research laboratory, CERN, in Geneva.

an axis. Just like tennis or billiard balls, they can have *spin*, but there is an important difference between particles and tennis and billiard balls. The spin (or, more precisely, angular momentum, which is roughly mass × radius × velocity of rotation) can be measured as being some multiple of Planck's constant divided by 2π. In terms of this unit, according to quantum mechanics, any object's spin has to be either an integer or an integer plus one-half. For each particle species the total spin – though not the direction of the spin – is fixed.

The electron, for instance, has spin $\frac{1}{2}$. This was discovered by two Dutch graduate students, Samuel Goudsmit (1902–1978) and George Uhlenbeck (1900–1988), who jointly wrote their theses on this subject in 1927. It had been an audacious idea that particles as small as electrons could have spin and, indeed, quite a lot of it. At first, the idea was met with skepticism because the 'surface of the electron' would have to move 137 times as fast as the speed of light. Nowadays such objections are simply ignored. There is no such thing as the surface of an electron.

Both photons and neutrinos, being massless particles, share the property that their rotation axis is always parallel to their direction of motion. Other particles rotate in arbitrary directions. It will, by the way, always be difficult to describe spin in ordinary words. Quantum mechanics makes it impossible to define accurately the direction of the rotation axis, except for the special statements made above. For very large, faster rotating objects, the direction of rotation can be given a more precise meaning.

Particles with integer spin are called 'bosons'; those with spin integer plus one-half are called 'fermions'. By consulting the spin values in Table 1, you can verify that the particles we call 'leptons' and the ones we call 'baryons' are fermions, and the mesons and the photon are bosons. In many respects, fermions behave quite differently from bosons. Fermions have the property that they each demand their own little space. Two fermions of the same type may never sit at the same spot, and their motion is always dictated by equations such that they continue to avoid each other. Curiously, no force is needed to accomplish this. There can be attractive or repulsive forces between fermions. The phenomenon that fermions each have to sit in a different 'state' is called *Pauli's exclusion principle*. Electrons are fermions (spin $\frac{1}{2}$). Now, every atom is surrounded by a cloud of electrons. If two atoms come too close to each other, the electrons will move in such a way that the two clouds avoid each other. The result is a repulsive force. If you clap your hands, you will notice that your two hands will not go through each other. This is due to Pauli's exclusion principle for the electrons in your hands.

By contrast with the individualistic fermions, bosons behave collectively. They like to sit all in the same place. A *laser*, for instance, produces a beam of light in which very many photons all share the same wavelength and direction of motion. This is because photons are bosons. We shall come across this collective character of integer spin particles again later.

There is another rule of the game our family of elementary particles must obey: for every particle there is a corresponding *antiparticle*. Particles have the same spin and exactly the same mass as their antiparticles, but the electric charges, as well as the numbers called S, I_3, L and B in Table 1 (I shall explain these shortly) are all opposite. For instance, π^+ and π^- are each other's antiparticles, just like K^+ and K^-. On the other hand, Σ^+ and Σ^- are not each other's antiparticles (they both have $B = 1$ and also their masses are not quite identical). The antiparticles of the latter have not been mentioned explicitly (see footnote [a] to the table). The particles π^0, η, and the photon, γ, are exceptions to this rule in the sense that these are identical to their own antiparticles.

Just like plants and animals, the observed particle types were sorted into species and families. Apart from the photon, we have *leptons* and *hadrons*. The latter are subdivided into *mesons* and *baryons*. This partition is based on the various kinds of *forces* occurring between these particles. The three kinds of 'forces' we will encounter are the 'strong force', the 'electromagnetic force' and the 'weak force'. I should add to this that when we talk of a 'force' this does not have to be something affecting the motion of these particles. Whenever particles affect each other in whatever way, including if they change each other's *identity*, we speak of a 'force'. Particles may exert forces on each other at a distance, but then this happens because they exchange a particle as a sort of messenger. These messengers will be called 'carriers' of the force. I must admit that at this point all of this must sound rather mysterious. In mathematical language it can all be described in a better way, a lamentation I will utter often. What I just described was the effects of a set of mathematical equations. Taken together, the equations actually make much more sense than my pedestrian English.

Back now to the particle table. The leptons (Greek $\lambda\epsilon\pi\tau\acute{o}\varsigma$ = light, skinny, feeble) are particles not sensitive to the strong force. They experience the weak force, and the electrically charged ones also feel the electromagnetic force. They are lighter than most other particles (although later a heavier member of this family would also be found), and, as far as we know, they all have spin $\frac{1}{2}$. This means that they do rotate about an axis, but at the smallest possible rate.

The lepton most familiar to us is the electron. Because this is the lightest electrically charged particle, all heavy, positively charged atomic nuclei more or less automatically collect a sufficient number of these electrons such that their

The weak force

The weak force is responsible for the fact that many particles and also many exotic atomic nuclei are unstable. The weak force can cause one particle to change into another, related, particle while emitting an electron and a neutrino. A general formula for the weak force was established by Enrico Fermi in 1934, and was later made more precise by George Sudarshan, Robert Marshak, Murray Gell-Mann, Richard Feynman and others. The improved formula worked well, but it was evident that it was not correct in all circumstances.

By 1970, only the first three of the following characteristics of the weak force were known to exist:

- The force acts in a universal way for many different types of particles, and its strength is nearly equal for all particles (although the *effects* of the force can be very different in different cases). Neutrinos are *exclusively* sensitive to the weak force.
- The force has a very short range, compared with any of the other forces.
- The force is very weak. Consequently particle collisions in which neutrinos are involved are so infrequent that very intense neutrino beams are needed if one wishes to study such events.
- The 'carriers' of the weak force are called W^+ and W^-, and would not be detected until the 1980s. Just like the photon, they have spin 1, but they are electrically charged and very heavy (which is why the range of the force is short). A third carrier, Z^0, is responsible for a different type of weak force that has nothing to do with particle decays: the 'neutral current'. It allows neutrinos to collide against other particles without changing their identity.

After 1970, the relationship between electromagnetism and the weak force became clear.

The strong force

The strong force acts only between the particles that are called *hadrons*. It provides these particles with a complicated internal structure.

Until approximately 1972 only the symmetry rules of the strong force were known, but we were unable to formulate the force laws accurately.

- The range of this force is not beyond the radius of a light atomic nucleus (10^{-13} cm, approximately).
- The force is strong. Particles that can decay under the influence of the strong force do so very quickly. They are named 'resonances'. An example is the Δ-resonance, with an average lifetime of only 0.6×10^{-23} s. If two hadrons come together within a range of around 10^{-13} cm, a collision is extremely likely.

Up until around 1972, the pions, having zero spin, and masses of 135 to 140 MeV, were considered to be the carriers of the strong force. For instance, the strong attractive force between two protons is primarily caused by pion exchange. Nowadays this is said to be due to the fact that pions happen to be the lightest hadrons, but just like the other hadrons they are composed of 'quarks'. The strong force is now viewed as a mere side-effect of an even stronger force between the quarks. The carriers of this stronger force are the gluons, to be discussed in Chapter 13.

electric charges are neutralized. Therefore, electrons occur in huge quantities in ordinary matter. A metal, for instance, owes its ability to conduct electric currents to the fact that a large fraction of all its electrons can roam there freely.

Neutrinos are also leptons. They only interact with other particles via the weak force, and that is the reason why they are extremely difficult to detect. A neutrino can travel through thousands of stars and planets without slowing down or changing direction.

The *hadrons* (Greek ἁδρός = strong) feel the strong force. This makes them very sensitive to one another's presence. One could argue that they are much 'bigger' than the leptons. Two hadrons closer together than about one fermi (10^{-13} cm) will almost certainly affect each other's motion or interact in some other way. This is not at all true for the leptons. By 1970, it was clear enough that this should give the hadrons a complicated internal structure, by contrast with the leptons, which one could best view as being 'point-like'. A hadron is a kind of ball made from some mysterious material.

The subdivision of the hadrons into *mesons* and *baryons* (Greek $\beta\alpha\rho\acute{\upsilon}\varsigma$ = heavy) was originally based on their difference of mass: the mesons have a mass that is usually between that of the leptons and the baryons. But the mass does not tell us everything about the nature of these particles; it is better to look at the spin. If the spin is integer, then we have a meson; if it is integer plus one-half, then we have a baryon (or an antibaryon). Most essential, however, is that in all processes among elementary particles the number of baryons minus the number of antibaryons always stays constant. We say that the total 'baryon number' is conserved; the baryon number, called 'B' in Table 1, is unity for baryons; $B = -1$ for antibaryons, and $B = 0$ for mesons.

One can think of other kinds of 'charges' that remain conserved when hadrons collide. We then talk of a 'conservation law'. One such charge is 'strangeness', indicated by the letter S in Table 1. Most particles have one fixed value for S. If particles A and B collide, and after the collision have changed into C and D, then such a collision process is only observed if the value of S of particles A and B together is equal to that of C and D together. The same is true for the total *energy* of A and B being equal to that of C and D together, and for a quantity called 'momentum', which for each particle is defined as mass $\times$ velocity. Conservation laws such as the conservation of energy, conservation of momentum, and now also conservation of strangeness, always play a vital role in particle physics.

The term 'strangeness' is quite apt: 'ordinary' particles, such as the proton, the neutron and the pions, have a strangeness of zero. It was Gell-Mann who discovered that other particles, such as the kaons and lambda particles, could be given a number such that this quantity, when added up, remains conserved during all known collision processes, and he proposed the word 'strangeness'. Another such quantity exists that we call 'isospin', for which we use the symbol I_3. The term 'isospin' perhaps sounds mysterious; it originates from the mathematical nature of this conservation law that reminds one of conservation of rotational motion, or 'spin'. It is as if a proton and a neutron can both be considered as being the same particle, called 'nucleon', but rotating in opposite directions in some 'internal' space ('isospin space'). This may sound mysterious to you, but for a mathematician this analogy provides new insights into the symmetries of these particles. For you can also let the nucleon rotate about different axes in isospin space and so obtain two other conservation laws, that of I_1 and I_2, but I will not make any further attempts to explain these.

We included the numbers strangeness (S) and isopin (I_3) in Table 1. You will notice that they are *not* always conserved when a particle decays. This is because the weak force, responsible for most of the decays, does not respect these conservation laws. Isospin is also violated by the electromagnetic force.

6 Life and death

When we talk about a particle's lifetime we always mean its *average* lifetime. A particle that is not absolutely stable has, at every moment of its life, the same chance of decaying. Some particles live longer than others, but the *average* lifetime is a characteristic of any particle species.

One can also use the concept of 'half-life'. If we have a large number of identical particles, the half-life is the time it takes for one-half of all the particles to decay. The half-life is 0.693 times the average lifetime.

One glance at Table 1 shows that some particles have a much longer average lifetime than others. The lifetimes differ enormously. A neutron, for example, lives 10^{13} times longer than a sigma-plus, and the sigma-plus has more than 10^9 times as long a lifetime as the sigma-zero. But if you observe that the 'natural' time scale for an elementary particle (which is the time it takes for their quantum mechanical state, or wave function, to evolve or oscillate) is somewhere around 10^{-24} seconds, you can safely state that *all* these particles are pretty stable. In our professional jargon, they are all called 'stable particles'.

How is a particle's lifetime determined? Particles with long lifetimes, such as the neutron and the muon, have to be captured, preferably in great numbers, after which one registers the decays electronically. Particles with lifetimes around 10^{-10} to 10^{-8} seconds used to be registered in a *bubble chamber*; nowadays this happens more often in a *spark chamber*. A particle moving through a bubble chamber leaves a trace of little bubbles which can be photographed. A spark chamber contains several sets of large numbers of thin wires crossing each other, with an electrical voltage between them. A charged particle causes a series of discharges (sparks) between the wires it comes close to. These are then registered electronically. The advantage of this technique compared with the bubble chamber is that the signal can be sent straight into a computer.

An electrically neutral particle never leaves a trace directly, but if it ventures into any kind of interaction involving charged particles (either it hits atoms in the detector or it decays into other particles), then this of course can be registered.

Furthermore, one usually places the entire apparatus between the poles of a strong magnet. This bends a particle's trajectory, and from this one can measure how fast the particle is travelling. However, because the bending also depends on the mass of the particle it is sometimes useful to measure the velocity in a different way as well.

Most particles in a high energy experiment will not move much slower than the speed of light. During their short lifetimes they can still travel for several centimeters. From the average length of such traces, one can deduce the lifetime. Lifetimes between 10^{-13} and 10^{-20} seconds, however, are very difficult to measure directly, but they can be determined indirectly by measuring the forces by which particles can transform into others. These forces are responsible for the decay, and knowing them one can calculate the lifetime. Thus, with endless skill, experimentalists devised a whole arsenal of techniques to figure out as much as is possible about our particles' properties. In some of these procedures it was extremely difficult to reach high precision. The numbers you see in Table 1 are the accumulated results of many many man-years of precision measurements, and the information there represents the latest published data.

That most particles have an average lifetime of 10^{-8} seconds or so actually means that they are *extremely* stable! Their internal wave functions oscillate more than 10^{22} times/second. This is their 'natural heart beat', with which the lifetimes should be compared. These quantum waves can oscillate $10^{-8} \times 10^{22}$, which is 10^{14} or $100\,000\,000\,000\,000$, times before decaying in one way or another. We can safely say that the force responsible for such a decay is extremely *weak*. Just imagine a tuning fork that would stop vibrating only after $100\,000\,000\,000\,000$ oscillations due to a very tiny friction. Real tuning forks stop much sooner because for them the friction is much greater. The 'friction force' that ends the life of some of our unstable particles has become known as the 'weak force'.

The decay of the neutron can also be attributed to the weak force, even though the neutron lifetime is much longer (a quarter of an hour on average). Incidently, some radioactive atomic nuclei, also decaying by the weak force, can take millions or even billions of years to decay. This wide spread of lifetimes can be explained by considering the amount of energy released in the decay. Energy is stored in the masses of the particles, by way of Einstein's well-known formula $E = Mc^2$. A decay can only take place if the total mass of all decay products added together is less than the mass of the original particle. The difference turns into energy of motion. If this difference is large, the process can happen very quickly, but often the difference is so small that the decay can take minutes or even millions of years.† Thus, not only the strength of a force, but also the amount of energy available determines the speed at which a particle decays.

If the weak force did not exist, most of the particles in Table 1 would have been perfectly stable. The force responsible for the decays of π^0, η and Σ^0 particles, however, is the electromagnetic force. You will notice that these particles have a much shorter lifetime. Apparently, the electromagnetic force is quite a bit stronger than the weak force.

This whole hotchpotch of subatomic particles first appeared in the 1950s and '60s. 'Had I foreseen this I would have gone into botany', was how Enrico Fermi, the famous Italian physicist, reacted.

There was much more to come. By 1970, a large series of hadrons with much larger spin values were also known. But they can decay via the strong force. Their lifetimes are therefore very short (all in the realm of 10^{-23} seconds). It is usually rather easy to figure out whether a given hadron can decay by means of the strong force: simply list all particle combinations such that, in total, all their charges togeher match with those of the original particle (including S and I_3), while the total mass added up should be *less* than that of the original particle. In that case there will be no further impediment against a strong decay into this particle combination. The excess mass is turned into energy of motion.

If all conditions are met except those of S and I_3, decay is still possible, but the job then has to be done by the weak force, which does not respect those quantum numbers. Weak decays take much more time.

If a particle's lifetime is as short as 10^{-23} seconds, the decay process has an effect on the energy needed to produce the particle before it can decay. To explain this, let us again compare the particle with a tuning fork in a vibrating mode. If the 'friction force' that tends to terminate this vibration mode is strong, this can affect the way in which it oscillates. The pitch, or oscillation frequency, is less well defined. For an elementary particle, this frequency corresponds to its energy. The tuning fork will resonate less precisely; its *resonance curve* becomes less peaked. Since similar curves are measured for these extremely unstable particles they are often called *resonances*. Their lifetimes can be deduced directly from the shapes of their resonance curves.

A typical example of a resonance is the *delta* (Δ), of which there are four species: Δ^-, Δ^0, Δ^+ and Δ^{++} (the latter has a double electric charge). The

---

† A decaying neutron has only 0.7 MeV of excess mass-energy that it can use to set into motion a proton, an electron and a neutrino. A radioactive nucleus usually has much less energy at its disposal.

masses of the deltas are nearly equal: about 1230 MeV. Their strangeness S is zero, and I_3 runs from $-1\frac{1}{2}$ to $+1\frac{1}{2}$; baryon number $B = 1$. All this implies that they can decay strongly into a proton or neutron and a pion, for instance:

$$\Delta^{++} \to p + \pi^+ \; ; \quad \Delta^0 \to p + \pi^- \; ; \quad \text{or} \quad n + \pi^0 \; .$$

On adding the masses, we see that there is a surplus of about 150 MeV. This is turned into energy of motion.

There exist both mesonic and baryonic resonances. The delta resonances are baryonic (later we shall encounter the mesonic resonance rho, ρ). Around 1970, dozens of resonances were known, and all evidence indicated that there would probably be infinitely many of them, each successor in the series having a little bit more mass than the previous, but a lifetime so short that their instantaneous presence would become very hard to detect.

Resonances seem to be just some sort of excited versions of the stable hadrons. They are replicas, either rotating faster than normal or vibrating in some way or other. This is similar to what happens to a gong when it is struck. The gong emits sound, losing energy as it does so, and eventually stops vibrating. Similarly, a resonance terminates its existence by emitting pions, whereby it transforms into a more stable form of matter.

7 The crazy kaons

It is not my intention to annoy you, my dear reader, with extensive discussions of each particle type separately, but the kaons are an exception. Some decades ago, one could buy in every toy shop a material called 'silly putty'. It was a miraculous substance. You could mold it like clay and roll it into a ball. But then when you threw the ball it would bounce quite elastically, not at all the reaction you would expect of a ball of clay. If you left your ball alone for a while, it would very slowly lose its shape altogether, turning into something that would look more like a puddle of water than clay, let alone a bouncing ball. Yet it still felt like and could be molded like clay. For a physicist such substances are interesting. The physical explanation of the self-contradictory behavior of this substance must lie in the special structure of its molecules. I suspect that they all are little threads of a special length sticking together by continuously winding around each other, releasing again by winding around others.

Neutral kaons are even more crazy than silly putty. They come in two types, K_{Long} and K_{Short}, but you can also say that there are two species that we may call K^0 ('K-zero') and $\overline{K^0}$ ('anti-K-zero'). If you have a K^0 then there is a 50% chance that it is a K_{Long} and a 50% chance that it is a K_{Short}. The same is true if you know you have a $\overline{K^0}$. On the other hand, if you 'know' you have a K_{Long} in your hands, then there will be a 50% chance that it will behave like a K^0, and 50% that it will turn out to be a $\overline{K^0}$. Also K_{Short} will behave in 50% of the cases like a K^0 and 50% of the cases like $\overline{K^0}$. However, a K_{Long} will *never* behave like a K_{Short}, and a K^0 will never behave like a $\overline{K^0}$!

A reader not very familiar with the wonders of quantum mechanics will not understand one iota of what I have just claimed. This is why I call the kaons the silly putty of the elementary particles. Give me another chance to try to illustrate further the miraculous statements I have just made.

A K_{Short} has an *average* lifetime that is less than 10^{-10} seconds, as listed in Table 1. Some of them will live a little longer, some will not live as long, but after, say, 5×10^{-10} seconds practically all of the K_{Short} particles will have decayed.

Now, K_{Long} has an average lifetime that is so long that the probability for one of them to decay within 5×10^{-10} seconds is very small. Therefore, if you have a neutral K meson that has decayed within 5×10^{-10} seconds you can be almost sure it was a K_{Short}. If it is still alive, you may safely assume that it is a K_{Long}, and you may expect it to decay only after some 5×10^{-8} seconds.

The quantum numbers S and I_3 do not apply to K_{Long} and K_{Short}, but if you produce a kaon via the reaction

$$p + p \rightarrow \Sigma^+ + p + K^0,$$

then you can be sure that the strangeness number S must be $+1$, because Σ^+ has $S = -1$. Since K^0 has $S = 1$ and $\overline{K^0}$ has $S = -1$ this reaction *must* have produced a K^0 and certainly not a $\overline{K^0}$. If we let a beam of neutral kaons run through a thin slice of material (a 'target'), the kaons can undergo all sorts of interactions there. But $\overline{K^0}$ are removed easier than the K^0, because the former can interact with nucleons in reactions such as

$$\overline{K^0} + p \rightarrow \Sigma^+ + \pi^0,$$

which K^0 cannot do, because all strange baryons have negative strangeness. Therefore K^0 can traverse a sheet of ordinary material much easier than its antiparticle $\overline{K^0}$.

Figure 7 shows what can happen to a beam of kaons if it is left on its own during some 10^{-8} seconds and then has to pass through a thin target. The particles produced, and the particles that survive the target are (predominantly) K^0. The particles that have not decayed after 10^{-8} seconds are surely K_{Long}, not K_{Short}. You can put any number of targets in the beam's way. Every time, the target absorbs 50% of the particles, and every time 50% of them decay immediately behind the target. Of course this sketch is an idealization. In a real experiment the kaons fly in all directions.

As you can imagine, this curious behavior gave rise to confusion when these particles were first discovered; it was thought at first that there were many different species of particles. In fact, we are dealing here with a typically quantum mechanical phenomenon. The strong force preserves strangeness, so, where a particle is produced, one always knows whether it is a K^0 or a $\overline{K^0}$. But after 10^{-10} seconds (and this is much *longer* than the duration of the production process itself, about 10^{-24} seconds), the weak force will be felt. Now the weak force totally ignores S and I_3. *Transitions* take place of the type

$$K^0 \leftrightarrow \overline{K^0},$$

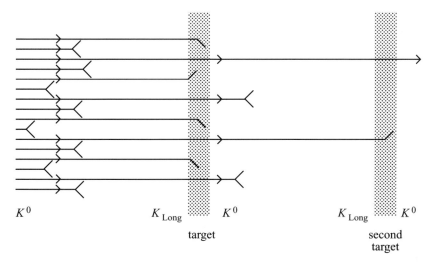

K^0 K_{Long} K^0 K_{Long} K^0

target   second
                                       target

Figure 7. Idealized rendition of an experiment with neutral kaons. After their production (left) approximately half of them decay quickly into two pions. The others are much more stable. We now know that these are K_{Long}. Of these, about half are absorbed by a thin target. the others seem to be immune; these must surely be K^0 particles. In turn, half of these decay quickly, and the survivors live much longer. One can repeat the cycle until all kaons are gone.

and this hopping system can occur in two different modes. This gives us the 'quantum states' K_{Long} and K_{Short}.

What happens is that both K^0 and $\overline{K^0}$ can decay into two pions, but 'interference' takes place. If the two sources, K^0 and $\overline{K^0}$, march in pace there is positive interference. The pion waves reinforce each other and the particle quickly decays. But if the interference is destructive the system cannot decay into two pions. We then identify it as a K_{Long}. The chances are always 50 : 50. K_{Long} *can* decay into *three* pions, but that takes much more time. End of explanation.

Alas, quantum mechanics is a difficult subject, and actually I do not expect my 'explanation' to be fully satisfactory. Never mind, you can also enjoy playing with silly putty without understanding how it works. It is not at all my intention to give the reader a feeling of being 'stupid', just so as to appear very learned myself. Quantum mechanics is a specialty not only requiring intensive study but also some 'getting used to'. I warned you about all this in my Foreword.

There is more fun to be had with the neutral kaons. To explain this I have to say more about the weak force. This force, as we have seen, fails to obey

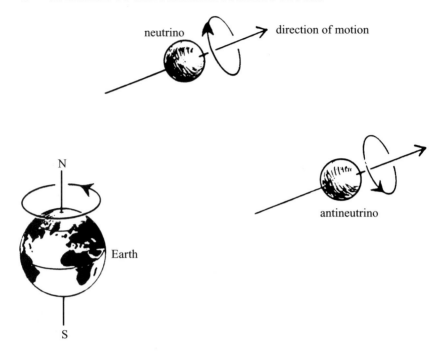

Figure 8. Direction of motion.

a certain number of 'laws of physics' such as conservation of strangeness and isospin. But there are other conservation laws that it does respect.

We shall often speak of *conservation of symmetry*. A very important, yet simple, symmetry is 'mirror-symmetry', officially called 'parity'. To see this, we compare particles with their mirror images. Before 1956, it was always assumed that any natural phenomenon respects the laws of Nature just as much as its mirror image. Consequently one expected that if particles or beams of particles hit each other in a mirror-symmetric way, the mirror symmetry would be preserved.

The discovery that many particles are not at all like their mirror images was made by two Chinese physicists, Tsung Dao Lee and Chen Ning Yang, some time after they had emigrated to the United States. It turned out that the weak force distinguishes between left and right. This is most obvious in the case of the neutrino. Neutrinos, v_e and v_μ, have no rest mass, like the photon, and so they always move with the speed of light. Neutrinos also rotate – they have spin $\frac{1}{2}$. Let us define 'north pole' and 'south pole' just like they are for the rotating Earth, see Figure 8. Neutrinos are special in always having their south poles in front of them and their north poles behind them. Neutrinos for which this is not the case have never been observed.

The Swedish physicist Cecilia Jarlskog compared neutrinos with vampires: they have no mirror image. Their mirror image is a physical impossibility. Now, I must apologize that my knowledge of vampirology is limited. Perhaps, if you are a vampire, you can still very vaguely recognize something like a mirror image. It could be like this for neutrinos as well; maybe they do have a mirror image that we have not yet seen. I will return to the wrongly rotating neutrinos in Chapter 19. What we do know is that *antineutrinos* have their north poles in front and their south poles behind. These particles are indicated as $\bar{v}_e$ or $\bar{v}_\mu$. It turns out that all particles resemble to a large extent the mirror images of their *antiparticles*.

The decay of K_{Short} into two pions is exactly mirror symmetric. This seemed to be difficult to reconcile with the kaon production process, unless there was a force during the decay that disturbed mirror symmetry. The argument here is again a somewhat difficult quantum mechanical story. If one tries to calculate the mirror image of the K^0 decay (or that of $\overline{K^0}$) into two pions, one discovers that by destructive interference this two-pion decay should completely be extinguished, *unless* the mirror image is different. In short, the fact that kaons can decay into two pions at all implies that the weak force has something of a corkscrew in it: that it differs from its mirror image. Now, we have already seen that the emission of pions by K^0 can be extinguished, by interference, by a similar decay of $\overline{K^0}$. This is how one can conclude that this particle resembles to a large extent the mirror image of its antiparticle. We call this 'PC symmetry', where P stands for parity, and C stands for the replacement of any particle by its antiparticle (conjugation). Only the product PC is a symmetry of the weak force; neither P nor C separately have this property.

But something else seems to be wrong. We have just stated that PC symmetry is the reason why destructive interference makes it impossible for K_{Long} to decay into two pions. James Christenson, James Cronin, Val Fitch and René Turlay decided to check this statement by doing an experiment. The sporadic K particles that, until then, had been studied in balloon flights high in the atmosphere, were not suitable for accurate measurements to be carried out. What was required were kaons produced in particle accelerators, so one could control the conditions much more precisely. (Val Fitch would later exclaim how much he regretted this; he much preferred working in the mountains to being under the ground.)

After very elaborate measurements and a lot of checking, these experimenters discovered that three out of 1000 K_{Long} particles *do* decay into two pions! What was so clever in this experiment is that it could be ascertained that there were no K_{Short} particles present, and, most of all, that no third particle was produced that could have escaped detection. Half a year of checking and further analysis were needed before the scientists were convinced. The interference phenomenon I mentioned earlier is not 100% watertight; there is a tiny leak!

What one should deduce from this is that the PC conservation law is also very slightly violated by the weak interactions. But you could also say that there is yet another force, a 'superweak force', whose sole purpose it is to sabotage PC symmetry conservation. Later we will see that such a force can be accommodated in our description of Nature, but the deeper reason for the existence of this force will remain a mystery for the time being. In 1980, Cronin and Fitch, the leaders of the experiment, received the Nobel Prize for their discovery.

There is yet another symmetry related to P and C: *time reversal*, T. One may follow any natural phenomenon backwards in time, for instance, by showing a movie of it in reverse. If we could do this for the planets in the solar system, we would see a scene that still obeys Newton's equations of gravity. Suppose that in an elementary particle theory we not only take the mirror image (P), but also replace all particles with their antiparticles (C) and go backwards in time (T). We call this a PCT transformation. After this transformation all laws of microscopic physics *must* remain the same. This follows by fairly delicate arguments from our theories. We are simply unable to describe any force that would violate this symmetry. One may also decide to check this experimentally, for instance, by checking if particles and their antiparticles have exactly the same masses. Deviations have never been detected, in spite of extensive searches by experimentalists, who would have loved to make fools out of the theoreticians.

The discovery of PC violation has often been connected to a problem in cosmology, the 'science of universes'. The enormous number of celestial bodies seen in our universe all consist of protons, neutrons and electrons. Astronomers are almost certain that there exists no single star or galaxy that is made out of 'antimatter': antiprotons, antineutrons and positrons (the antiparticles of electrons). Whence this lack of symmetry? Well, maybe in the very first stages of the universe there was no difference at all between the amounts of matter and antimatter. All that existed was a hot 'soup' of extremely energetic primordial particles. After that, the forces of Nature could have acted on these primordial particles to condense out of these the particles of which we are all made. But somehow there must have been a slight preference for matter instead of antimatter. These forces would have had to violate the law of baryon number conservation, but that is not a problem; we will later encounter such forces.

However, these effects could never take place without violating PC symmetry, if we may assume that the universe by itself is approximately mirror symmetric.

Now that we know of the existence of a PC symmetry violating force, we can imagine a universe that started out small and without preference for matter against antimatter, and that evolved into the world in which we live. This can only have happened during a time when the universe was evolving rapidly, otherwise one could have used time reversal symmetry to deduce that the baryons would neatly balance out against the antibaryons. However, constructing models such that all details of such a mechanism also work out right is, as yet, a science uncomfortably closer to science fiction than science fact.

8 The invisible quarks

Once the order among the numerous particle species had been clarified, a certain pattern became recognizable. Just as Dmitri Ivanovich Mendeleev discovered the periodic system of the chemical elements in 1869, a similar system among the particles became visible. This pattern was found independently by the American Murray Gell-Mann and the Israeli Yuval Ne'eman. Eight meson species, each with the same amount of spin, or eight baryon species, each with the same spin, could be arranged beautifully into packages we call *multiplets*. The corresponding mathematical scheme is called $SU(3)$. Grouplets of eight pieces form a 'fundamental' octet. This is why Gell-Mann called his theory 'The Eightfold Way'. He had borrowed this from Buddhism, according to which the way to Nirvana is the Eightfold Way.

But $SU(3)$ mathematics also admits multiplets containing *ten* members. When this scheme was proposed *nine* baryons were known with spin $\frac{3}{2}$. (They do not figure in our Table 1 because they are resonances. Four of them are the Δ resonances mentioned earlier; they decay into stable nucleons and pions.) The $SU(3)$ patterns are obtained when two fundamental properties of the particles are plotted on graphs of strangeness S against isospin I_3. I have sketched a few of them in Table 2.

So Gell-Mann predicted a tenth baryon, the *Omega-minus* (Ω^-). He could predict pretty accurately its mass, because the masses of the other nine baryons turned out to vary in a systematic way within the plot (and he could understand these mass variations as being a consequence of a simple force). However, it was now clear that the Ω^- has nothing to decay into that would not be forbidden by the strong force conservation laws, because its strangeness S has the extreme value -3. So, the Ω^- can only decay weakly, and therefore its lifetime could not be as short as 10^{-23} seconds like the other members of this multiplet, but had to be something of the order of 10^{-10} seconds. Consequently, this particle can travel several centimeters before it decays, and that would make it easily detectable. In 1964 the Ω^- was found, with precisely the properties predicted by Gell-Mann.

Table 2. *Two octets and a decuplet*

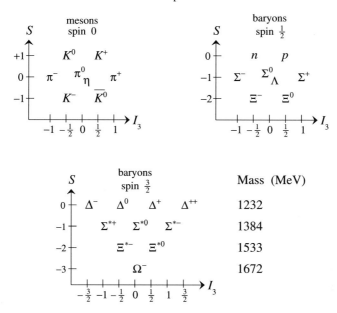

Multiplet structures were identified for most other baryons and mesons. Gell-Mann also realized how one could explain these structures. He suggested that the mesons, as well as the baryons, should be composed out of 'even more fundamental' building blocks. Gell-Mann worked in the California Institute of Technology in Pasadena, where he often conversed with Richard Feynman. Gell-Mann and Feynman are both famous physicists, but with quite different personalities. Gell-Mann, for example, is known as an enthusiastic bird-watcher, well at home in arts and literature, and proud of his knowledge of foreign languages. Feynman was much more the self-made man, a no-nonsense analyst, making fun of anything resembling established authority. There is an anecdote that seems to have no basis in fact, but sounded too good to me not to be told; it *could* have happened this way. Gell-Mann told Feynman that he had a problem. Here he was proposing a new kind of building block of matter: what should he call it? Undoubtedly he must have thought of using learned Latin or Greek terminology, as this had always been customary in scientific nomenclature. 'Nonsense', Feynman said. 'You are talking here of things never thought of before. Those beautiful but old-fashioned words are out of place. Why not just call them "shrumpfs" or "quacks" or something like that?'

Table 3. *The quark and antiquark triplets*

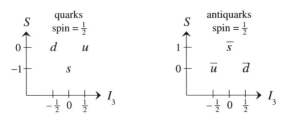

Table 4. *Quark–antiquark composition of mesons (the particles in the center can also temporarily behave as $s\bar{s}$)*

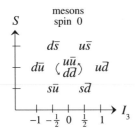

When I asked him later, Gell-Mann denied that such a conversation with Feynman ever took place. But quarks it would be. Gell-Mann's explanation was that the word comes from a phrase in James Joyce's *Finnegan's Wake*: 'Three quarks for Muster Mark!' And yes, that figures. These particles like to be with three together. All *baryons* are built with three quarks, whereas the *mesons* consist of one quark and one antiquark.

The quarks themselves form an even simpler $SU(3)$ pattern, as sketched in Table 3. We call them 'up' (u), 'down' (d), and 'strange' (s). 'Ordinary' particles only contain up and/or down quarks. 'Strange' hadrons contain one or more s quarks (or $\bar{s}$ antiquarks). Table 4 shows how the meson octet can be built from quarks and antiquarks.†

The quark composition of spin $\frac{3}{2}$ can be seen in Table 5. Why the spin $\frac{1}{2}$ baryons only form an octet is more dificult to explain. It is related to the fact

---

† With three quarks and three antiquarks, of course, *nine* combinations are possible, but the ninth, a state continuously changing from $u\bar{u}$ to $d\bar{d}$ and $s\bar{s}$, behaves exceptionally. This object, called η', is considerably heavier than the others.

Table 5. *Quark composition of the baryons*

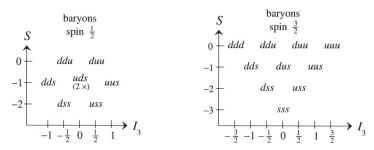

that in these states at least two of the quarks have to be different from each other.†

Actually, the idea of simple fundamental building blocks for hadrons had also been conceived by others. George Zweig, also at CalTech in Pasadena, had such an idea. He had called his building blocks 'aces'. But the word 'quarks' has stuck. Why some scientific names are more successful than others is sometimes difficult to trace.

But there were some odd aspects to this theory. Apparently, quarks (or aces) always exist in pairs or threes, and are never seen alone. Numerous attempts have made by experimentalists to detect isolated quarks in their specially designed apparatus, but none of them were successful.

And quarks, *if* they could be isolated, would exhibit even stranger properties. What, for instance, should their electric charges be? It is reasonable to assume that all *u* quarks always carry the same charge, and all *d* and *s* quarks should also always have the same charge. Compare now Table 5 with Table 2. Clearly these suggest that the *d* and the *s* quarks have electric charge $-\frac{1}{3}$, and the *u* quark has charge $+\frac{2}{3}$. But particles having a charge that is not an integer multiple of that of the electron or the proton have never been observed. If such particles existed it should be possible to detect them experimentally. That they have always escaped experimental detection must mean that the forces keeping them bound together inside a hadron must be unbelievably efficient.

---

† But we can understand this more precisely. Given three types of quark, each either with 'spin up' or with 'spin down', we have in total six kinds. The total number of different combinations of three of these can be counted to be fifty-six. Each element of the decuplet has spin $\frac{3}{2}$ and can therefore rotate in four different ways around its axis (one of those wondrous facts in quantum mechanics that I will not further elaborate on). The members of the octet have spin $\frac{1}{2}$ and therefore can rotate in only two different ways. And then the numbers work out: $56 = 4 \times 10 + 2 \times 8$.

With the advent of the quarks, the flora and fauna of the subatomic particles had become slightly clearer. But still they formed a weird lot, even though only a few of them occur in huge quantities in the universe (protons, neutrons, electrons, photons). And, as Sybren S. de Groot once exclaimed when he studied neutrinos, you really fell in love with them. My fellow students and I loved these particles. Their behavior was a big mystery. The leptons are the simplest, because they are point-like, but they have spin, and because of that the weak interaction could still act in a very complicated way on them. But the weak force had been pretty well documented by then.

The hadrons were much more mysterious. Collision processes between them were too complicated for any respectable theory. If you tried to imagine them as little spheres made of some sort of material, you were still left with the problem of how to understand quarks, and why they continued to resist all attempts by experimentalists to isolate them.

9 Fields or bootstraps?

There has existed a *phenomenological* theory for the weak force since 1958. That is to say, there existed formulae that correctly described the effects of the weak force for all particles, under all circumstances, that could be realized in the experiments of that time, with margins of error within a small percentage. The theory could never be a *fundamental* theory, because it was understood that the theory would break down in more violent collision experiments. But it was also known that it would take at least a decade for the experimentalists to reach such energies in a laboratory, and so it seemed that for the time being this 'temporary' theory would have to be sufficient.

For the *strong* force there were a hotchpotch of phenomenological theories. None of these were very accurate, and this situation was obviously very unsatisfactory. The only features of the strong interactions that were well understood were its various symmetry properties and the ensuing conservation laws. Strangeness, isospin and a few other quantities were very strictly conserved for this force.

This was in striking contrast to the situation with the *electromagnetic* force. This has the peculiarity that it can propagate over long distances, and so this force is also experienced in the everyday world. The British physicist James Clerk Maxwell had given the mathematical formulation of *electrodynamics* as early as 1873 (see Figure 6).

The ideal particle for studying the effects of electromagnetic forces is the electron. It is a lepton, which means that it is insensitive to the strong force. Its mass is much smaller than that of most other particles; only later would people realize that this is precisely the reason why all sorts of indirect effects due to other particles on the electron can be neglected to a good approximation. The weak force does affect the electron, but its effects are so weak here that it can safely be disregarded. The electrons circling around the nucleus are like a miniature planetary system, and it is exclusively the electromagnetic force that is in charge here.

It is understandable why the fundamental theory for the interactions between

electrons and photons was completed first. It was called 'quantum electrodynamics' or QED. The accuracy with which all sorts of properties of the electron
could be calculated using this theory was impressive. One of the most striking
examples of this was the calculation of the electron's *magnetic dipole moment*, μ.
Since the electron rotates about its axis and is also electrically charged, it acts
as a miniature magnet. Paul A. M. Dirac, who was the first to write down a
quantum mechanical equation for the electron that also agreed with relativity
theory, found that the strength of this little magnet could be calculated in terms
of known constants of Nature, these being Planck's constant, the speed of light
and the electric charge and mass of the electron:

$$\mu = \frac{eh}{4\pi m_e} \ .$$

But his equation still neglected certain indirect effects on electrons due to photons
surrounding them; these effects were later calculated by Julian Schwinger and
others with ever increasing accuracy. The experimentally measured value of the
electron magnetic dipole moment is presently

$$1.001\,159\,652\,19 \ \pm \ 0.000\,000\,000\,01$$

times the combination of constants given first by Dirac. Quantum electrodynamics gives for this number

$$1.001\,159\,652\,17 \ \pm \ 0.000\,000\,000\,03 \ .$$

As you can see, theory and experiment are competing to obtain the best possible
accuracy. The agreement is nearly perfect.† For comparison, if one could measure
the distance from here to the Moon with the same relative accuracy, the margin
of error would only be a few millimeters. The theory seems to run behind a bit:
it has a slightly greater margin of error, an inaccuracy of one centimeter, but this
is mainly to be blamed on the inaccuracies inherent in determining the charge of
the electron.

Before this brilliant result could be obtained however, some mountains had to
be moved. An essential aspect of elementary particle theory is that particles can
be 'created' and 'annihilated'. A consequence of this is that the total number of
particles involved in any interaction with an electron is continuously changing.
There are two ways of looking at this theory: on the one hand, we are used
to describing all processes in terms of particles being created and annihilated in

---

† These values are correct as of 1995.

various places and times; but we can also view these processes as an ocean of waves crossing and affecting each other. These waves are said to be waves in 'fields'. Every particle type has its corresponding field. The photon belongs to the electromagnetic field. The electron has a very curious type of 'electron field' (the 'Dirac field'). In their wave-like oscillations these fields react upon each other's presence, and all of this is neatly governed by their 'field equations'. Quantum electrodynamics was the first system for which these equations became known, and so became the first prototype of a 'quantum field theory'.

By 1930, physicists were already aware of the problems that had to be solved. Dirac realized the necessity to introduce the electron's antiparticle, the *positron*,† which was experimentally discovered by Carl D. Anderson in 1932. The theory of quantum electrodynamics was gradually improved by Paul Dirac, Julian Schwinger, Sin-Itiro Tomonaga, Richard Feynman, Freeman Dyson, and many others.

The first difficulty was that this theory could not be formulated in a mathematically perfect way, but only as a series of successive approximations, each approximation being more accurate than the previous ones, but none of them 'exactly' right. The biggest difficulty, which seemed to jeopardize the entire construction, was the fact that the first calculations of these 'tiny corrections', such as the ones needed to calculate the precise value of the electron's magnetic dipole moment, repeatedly turned out 'infinity' as an answer.

That, of course, is nonsensical; and by 1970, scientists knew precisely how to deal with this problem. Quantum electrodynamics is said to be a *renormalizable* theory. Roughly, this means the following (I will return to this on several occasions). We have to start with the so-called 'naked electron', that is, an electron without any photons in its vicinity. Actually, an electron cannot be separated from its surrounding photons, but we ignore this for the time being. This naked electron is given some value for its 'naked' electric charge and its 'naked' mass. If we try to calculate the magnetic moment for this naked electron we find, regrettably, that it is infinite. This is nonsense. But if we calculate what happens if photons enter the vicinity of this electron, we find that these photons create new electrons and positrons. This would not be noticed directly, but these additional particles have all sorts of effects on the electron. First, they act as a neutralizing shield against the electric charge. This effect is called *vacuum*

---

† Some linguists claim that the correct word should be *positon*, since this particle has a positive, not 'positrive' charge, in contrast to the negatively charged electron. The word, however, is derived from **posi**-tive elec-**tron**.

polarization, and it causes the electron's charge to change. If we try to calculate this change, alas, we find it to be infinite also. Secondly, the extra photons, electrons and positrons have an effect on the electron's mass (they carry energy, hence also mass). This mass change also turns out to be infinite. In short: the true charge and the true mass of a real, 'physical', electron, are very different from those of the naked electron.

If we want to formulate the theory very precisely we must realize that an experimenter never studies a naked electron. He† only observes the physical electron, and he only measures the total charge and mass of this physical electron. When considering the constants of Nature that we use to express the electron's magnetic moment, only the observed values should be used, and not the values for the naked electron. In summary, we should compare the electron magnetic moment directly with the charge and mass of the physical electron, the one that is really observed. The replacement of the 'naked' charge and mass of the electron by the physically observed values is called 'renormalization'. If we obey the rules following from this carefully enough, we find, somewhat surprisingly, that all the annoying infinities cancel out. A useful expression for the strength of the electron magnetic dipole moment remains, and it yields the number mentioned earlier. Ergo: the infinities are not in the electron or in the forces acting on the electron, but solely in our hypothetical naked electron. The naked mass and the naked charge are infinite (or, rather, ill-defined), but we can never observe these anyhow.

By no means everybody was happy with this reasoning, but the theory worked so well that the protesters were outvoiced by the enthusiasts. If a theory even remotely as successful could be composed for the other forces of Nature, we would be more than content. Most of the remainder of this book will recount how this was exactly what was about to happen, to the surprise of many.

It did not look that way, however, in 1970. The cheering about the successes of quantum electrodynamics had subsided a bit. The calculation of the value of the next decimal place in the electron's magnetic moment was becoming viciously complicated and of relatively little interest. The weak force, as far as it was understood, would certainly not be renormalizable. It was relatively easy to convince oneself by looking at the formulae that one should not even try to compute the effects of vacuum polarization arising from the weak force, because the infinities could not possibly cancel out. All those particles that would be

---

† The reader should note that by 'he' I always mean 'he or she'. Regrettably, the number of female theoretical physicists was, and still is, very small.

created by the weak force to polarize the vacuum would simply have to be ignored. Things would come out approximately, but not exactly, right.

As for the strong force, maybe one could design a renormalizable theory for it, but this would not be of much help, because successive approximations would continue to give different answers. This happens because the force is so strong that every 'correction'would be bigger rather than smaller than the previous one. This situation can be compared with that of a golf player who is unable to strike the ball softly. Each stroke brings the ball further away from the putt.

'We have to do this differently', is what most investigators thought. And they had many reasons for thinking this. In a renormalizable theory, where one works with successive approximations, the naked particles must, by first approximation, move freely in straight lines. But what if the naked particles are quarks? They cannot move freely at all! You cannot isolate them, or at least you cannot observe them in isolation. Maybe quarks are not particles at all, but resemble ghosts. 'They are mathematical objects, not real particles; you have to set up such a theory in an entirely different manner,' was the common census of opinion. And so alternative theories were devised: some would be quite helpful in searching for the ultimate solution to these questions; others were totally off-target.

Actually, this was a situation that had been encountered before in the development of scientific theories, and it would emerge again later. The problems could not be solved by exact mathematical techniques and analyses alone, or at least not with the knowledge available at the time. The only way to make progress was to try out all sorts of intuitive ideas. The existing colorful set of observed particle types, as well as the multitude of poorly understood experimental data about them, presented a severe test to human ingenuity. Would any comprehensive theoretical understanding of these particles be possible at all? No one could tell. But there was no shortage of ideas, because, fortunately, our world is also populated with a colorful crowd of physicists.

First we had the *formal mathematical* approach, propagated mostly by those who were sick and tired of the renormalizable field theories with their shaky mathematical justification, and which did not seem to apply to most of the particles and forces that had been registered so far. According to this reasoning, the only things one was allowed to use were, first, the set of particles existing *before* a collision process takes place and the *quantum states* that they can be in (the 'in'-states), and then the set of particles and states that were observed *after* the collision, that is, the 'out'-states. Not every out-state can be reached from every in-state. The laws of quantum mechanics provide all sorts of restrictions. One of these restrictions is that the out-state may never materialize before the in-state has been realized. We call this 'causality', the logical order of cause and effect. The mathematical concept that was put at the center of this approach was the *S-matrix*.†

It was realized that there could well be an infinite sequence of particle types (not only those in Table 1, but also the presumably infinite series of resonances, the short-lived subatomic fragments). In this formalism there was no reason to call some of these particles 'elementary' and others 'composites'. For instance, the delta resonance ('Δ') can decay into a proton and a pion. Alternatively, a proton, if colliding against something else such that its energy is increased, can decay into a delta and several pions. Should one say then that the delta is composed of a proton and a pion? Or is it the other way round, does the proton consist of a delta and a pion? If all you are interested in is the *S*-matrix, such a question is irrelevant.

This idea was called the 'bootstrap theory', after the mythological figure who tried to lift himself up by pulling at his own bootstraps. One of the pioneers of this theory would later motivate it by using a rather dubious version of 'holism': that the particle family is one big system which you should not try to reduce further into more basic subunits.

Whether they were along the right lines or not, these investigations did result in a host of mathematical schemes and properties that were useful in understanding the *S*-matrix. One of the pioneers of *S*-matrix theory was my uncle, Nico van Kampen. Expanding the ideas of two other Dutch physicists, Hans Kramers and Ralph Kronig, van Kampen deduced in the early 1950s that if an *S*-matrix obeys causality and also the laws of relativity, certain equations follow that are called *dispersion relations*. Having a sharp mind, a vast knowledge of theoretical physics and a rock-solid dedication to his work, van Kampen is one of Holland's most respected theoretical physictsts, and a continual inspiration to me. As a student, I would frequently burst into his office with my immature theories about anything related to theoretical physics. His usual reaction was to reformulate the questions I had used as starting points, and, more often than not, my beautiful ideas quickly evaporated.

Van Kampen was irritated by the unscrupulous way in which dispersion equations were being abused to 'prove' properties of particles that could certainly not be understood in that way. He turned away from this subject, and eventually

---

† 'S' stands for scattering, the collision process among particles. The states are seen as mathematical vectors. A matrix is a square table of numbers with which vectors can be transformed into other vectors.

it would turn out that *S*-matrix theory alone would be insufficient to understand particle properties quantitatively. The rich symmetry pattern could not be deduced, and the quarks remained a mystery.

If one really wanted to understand the structure of the fundamental particles, a further ingredient had to be included in the mathematical theories: the observed *symmetry structure* and the associated *conservation laws* that are obeyed by the particles. As a rule of thumb, one can state that every conservation law (conservation of energy, of strangeness, baryon number or whatever) always corresponds to a 'symmetry' in the system of particles. This rule had been discovered in 1918 as a property of field equations by the eminent mathematician Emmy Noether (1882–1935).

If an electric charge (an example of a conserved quantity) moves from one place to another, we experience something we call *electric current*. All the other conserved quantities, such as strangeness and isospin, should be able to produce similar 'currents', and one could try to describe these currents accurately. In combination with the *S*-matrix theory, this led to something that was called 'current algebra'. Current algebra soon produced interesting new insights. One example is the mystery resulting from the decay of charged pions. These particles nearly always decay into a muon and an (anti-)muon-neutrino, almost never into an electron (or positron) with a corresponding electron-neutrino. But why? After all, the electron looks as much like a muon as if it were its (lighter) twin brother.

The reason for this preference for muons in the pion decay turns out to be that this decay can be attributed completely to a kind of current that occurs within the pion (the so-called axial vector current), which in turn produces a current of electrons, muons and neutrinos. The mathematical expressions tell us exactly how this comes about. When a pion decays into leptons, the axial vector current is switched off and the leptonic current switches on. The latter can be interpreted as a shower of lepton pairs. One can now calculate precisely the probability that a particular lepton pair with the allowed spin orientations will be produced; it turns out that this probability is proportional to the square of the charged lepton's mass. The muon is about 200 times as heavy as the electron, and one finds therefore that the efficiency of a pion decaying into a muon and its antineutrino is $200 \times 200 = 40\,000$ times as large as that of the electron decay.

Many of these sorts of relations had been found in the particle properties that could be explained using current algebra. In particular, most of the weak decays could be calculated accurately by assuming that the currents inside all decaying particles are essentially the same. Nevertheless, it seemed that not everything the particles do could be explained in this way. As stated earlier, the hadrons behave more or less like billiard balls that can collide against each other. How large and how sturdy are these billiard balls? How strongly do they attract or repel each other? It seemed to be impossible to answer all such questions. Should one return to the field theories after all?

The theory for the weak force that we had was the one originally designed by Enrico Fermi in 1934. He had written down the most general expression that can yield the kind of transitions one sees in the weak interaction processes. The formula still allowed for a lot of different possibilities, which could be excluded one by one by comparing them with experimental data. Finally (after some misunderstandings and false claims had been dealt with), only one mathematical expression remained. It had been discovered by George Sudarshan, Robert Marshak, and independently by Richard Feynman and Murray Gell-Mann. It implied that a certain kind of current that can be found in all particles produced another current *at the same spot*, and that this other current produced the particles into which the previous ones may decay. Unfortunately, this so-called 'current-current theory' was surely not renormalizable. If a current causes another current to arise at the same spot only, you can view this as resulting from a force much like electromagnetism, but with an extremely short range of action. Vacuum polarization certainly does take place in such a theory, but it adds to the naked particle forces that are *different* from the original ones and yet infinite. Now, we do not know how to redefine the properties of the naked particles in such a way that the infinities cancel out. Essentially what this means is that if you want your theory to produce finite, meaningful results, and if you want to calculate the necessary corrections, you have to add new and as yet unknown forces to the existing ones. This adds more and more unknown features to a calculation and results in making the final answer completely void: there is just no way to tell how big all these corrective forces are.

Some investigators then made an error that would be repeated many times subsequently (even up to the present day!): they thought that there were short-comings in the mathematical methods. By writing the equations in a different way, putting the corrective terms in a different order, and so on, they hoped to improve the theory in such a way that it would spit out sensible results. It was thought, erroneously, that mathematics could turn wrong equations into right equations; but this can never be the case. There simply was no reason to assume that we were starting with anything that looked like the true equations.

And so it happened that only one last straw remained to hang on to: the search for a renormalizable field theory for the weak force, in which one starts off with a small number of 'fundamental' fields, from which other fields can be constructed by combining some of them together. One may also say that a number of 'fundamental' particles are postulated, from which all other particles can be built.

Most particle physicists considered such ideas as being outmoded. Only a handful of diehards were seriously fiddling with these ideas. The renormalization procedure was considered by most to be artificial and ugly. Why are some particles fundamental and others not? And, what is more, we did not want a theory in which calculations could only be made by making successive approximations, in which infinite forces balance each other out. The exact theory, all at once, is what we wanted.

If we now look back at papers published in the 1960s, it strikes us that whoever talked or wrote about renormalization usually began their arguments with extensive apologies: it was of course known that renormalization was probably a blind alley, but, who knows, maybe these ideas could eventually be good for something.

CERN, the European Center for (sub-)nuclear reasearch near Geneva, does not only consist of immense laboratories, allowing physicists to look deeply into the constituents of matter, there is also a theory division. There were theoreticians there who, being very close to the heat of the experiments, realized that infinite forces cancelling each other out were a reality that could not be avoided if one wanted to perform detailed calculations that had to be performed routinely in order to understand the experiments.

In 1969, Sheldon L. Glashow (we will meet him again), John Iliopoulos and Luciano Maiani, at CERN, published a paper, the importance of which was not immediately recognized, but which would play a dominant role in what was going to happen. They noticed that if they introduced next to the known quarks – *up*, *down* and *strange* – a fourth, the infinite forces seemed to cancel each other out much better than before (although the theory was still unrenormalizable). Such a fourth quark had been suggested earlier by Glashow and James D. Bjorken. They had found the resulting symmetry pattern so charming that they dubbed the newly proposed quark '*charm*'. Iliopoulos and Maiani were happy to adopt this name; 'charming' after all also means 'enchanting'. As if by magic, infinite forces canceled other out.

But the GIM mechanism, as it was to be called, did not really imply a new renormalizable theory for the weak interactions. There was one person who was completely determined to figure out how to construct a fully renormalizable theory. Martinus Veltman, who had recently been appointed as professor of theoretical physics at the University of Utrecht, was wary of abstract mathematical philosophies. He wanted a pragmatic approach to figure out what happened, in reality, in these elementary particles. Before he came to Utrecht, he claimed, nobody there knew what a kaon was. This was going to change. Veltman frequently went to CERN and Paris to learn and discuss. Why do these infinite forces cancel out so nicely? Why, in spite of the infinite forces, do the effects of the weak force satisfy so neatly the rules of current algebra? Why does the muon look so much like the electron? How precisely does the GIM mechanism work? Veltman went into deep discussions with John Bell at CERN to weigh the various arguments against each other.

Once I had passed my examinations at Utrecht, I wanted to continue as a graduate student to study elementary particles. My uncle had turned to a different subject in theoretical physics, and Veltman became my teacher.

Let me introduce Veltman with a little anecdote, which shows him to be at home in both the theory of gravity as well as in the intricacies of modern technique. This he demonstrated on an occasion when he was one of the last persons to enter an elevator that was already loaded with people. When the button was pressed, a buzzer sounded and a signal flashed: overloaded! Since Veltman was the heaviest person in the elevator, and also one of the last to enter, all eyes fell on him. But Veltman did not agree that he should step out. 'When I say "yes" then press!', he said. He bent his knees and then jumped, higher than was to be expected for a person of his stature. 'YES!' he yelled, and the elevator took off. By the time he came down the engine had apparently gained enough speed for the elevator to continue its journey.†

Anyway, I began to study field theory. People told Veltman that he and his students were dusting an old and deserted corner of physics. How dusty and deserted you will read in the following chapters.

† I was in that elevator.

10 The Yang–Mills bonanza

We must now go back to 1954, to a time when the big successes of the theory for the electromagnetic forces between the particles were still fresh. Not yet discouraged by the discoveries of numerous particle species that were yet to come, scientists were still searching for simple, elegant and universal principles in physics.

All aspects of the electromagnetic forces between particles can be deduced from the equations that are obeyed by the *electric* and the *magnetic* fields. These fields are vector fields. A 'vector' is a quantity that is not only characterized by its *strength* but also by its *direction*, and several numbers (typically three) are needed its description. 'Field' means that these numbers may have different values at different points in space and time. The wind velocity in the atmosphere could be called a vector field. For the wind, you first indicate how much air is moving northwards (if it moves southwards you give this number a minus sign); secondly, you determine the east–west movement; and finally you give a number corresponding to the vertical component.

Taken all together the electromagnetic field has six components. But these cannot be chosen at will. These numbers are hanging together via 'field equations', just like the wind velocity components are related to the air pressure distribution in the neighborhood. If you know the air pressure, you can calculate the wind speed in all directions. We have something like this for the electromagnetic field as well: you can introduce the electric *potential* field, which, just like air pressure, is given by one single number, not three. But if the fields are time-dependent, and if there are magnetic fields, you also need a *vector potential field*. So all together you have four potential fields. Now, mathematicians can work with vectors having four components just as easily as they can with those with three components. So, we take these potential fields together and speak of the so-called '4-vector potential'. If you know the 4-vector potential everywhere you can deduce the six electric and magnetic field components.

There is, however, something odd about the 4-vector potential. It is not directly observable. To say this more precisely, many different 4-vector potential fields can be invented that all generate the *same* electromagnetic fields everywhere, and therefore they would be indistinguishable to an experimenter. For example, the Earth could be given a voltage of 100 000 volts, and two holes in my electricity outlet could be carrying 100 000 volts and 100 120 volts, respectively. My 120 volt (or 240 volt) dishwasher will function totally normally, and I will not be electrocuted by it.

Now, you might suggest that such an arbitrariness would render the 4-vector potential practically useless, but, in fact, the opposite is true. The equations of motion for the elementary particles become a lot easier and more elegant if this 4-vector potential is used. The arbitrariness, which we call *gauge invariance*, does no harm at all.

In practice, gauge invariance is actually quite convenient. If, for example, you perform a complicated calculation and you want to assure yourself that you have not made any kind of mistake, gauge invariance is a nice way to check everything: you simply change the 4-vector potential in such an invisible manner. We call this a *gauge transformation*. The final result (which particles have been produced, how many and in which direction) should not change at all. Or else it could be that we altered the *names* of these particles, but they should still be indistinguishable from the previous ones in an experiment. Using the example of the dishwasher, if I calculate how much noise it makes, the result should not change if I add 100 000 volts to both holes in my electrical outlet.

Gauge invariance is a key issue in the theory of quantum electrodynamics. A similar principle can also be found in Einstein's theory for gravitation. There it is the coordinates of points in space and time that may be chosen at will. But the physical phenomena as observed by someone who does an experiment should not depend on how these labels were attached to points in space and time.

As gauge invariance was such an important principle in the only two forces that were well understood in 1954, it was natural to try to construct similar 'gauge theories' for the other forces. This was exactly the starting point in a very elegant calculation presented in that year by Chen Ning Yang together with his younger collaborator Robert Mills. As it turned out later, others had had the same idea. Ronald Shaw, a student of Abdus Salam's in Cambridge, England, had almost completed his thesis when the Yang–Mills paper appeared. The results he had described in one chapter of his thesis were practically the same, but were never published. Apparently Shaw accepted that someone had beaten him. Scientists of our time are often not such good losers; in such circumstances, they often try to have their work published quickly anyway, just to be able to claim later that they made their discovery 'independently'. Just like soccer players, scientists seem to have become more aggressive these days.

What Yang and Mills had proposed was to extend the set of possible gauge transformations. This is made possible if more components to the 4-vector potential field are added. The first interesting case is a potential field with *twelve* rather than four components. In such a world, there are three kinds of electric and three kinds of magnetic fields, the *Yang–Mills fields*.

It was hard to construct sound arguments in favor of what Yang and Mills had proposed. Their work should not be regarded as an attempt to explain anything about the behavior of the particles that had been observed until that time. What their calculations referred to could only be seen as a 'dream world', with an unrealistic simplicity and abstraction. This is a kind of exercise that is becoming more and more habitual in theoretical physics, and it should be seen as a way to exercise and sharpen our mathematical machinery. We should then talk of a 'model'. In contrast, we should really reserve the word 'theory' for all those cases when a model is asserted to be a (possibly idealized) description of the real world. Regrettably, the words 'model' and 'theory' are very often mixed up in modern publications. Authors often prefer to call what they are working on a 'theory', even if they have not made the slightest attempt to figure out what it refers to in the real world.

Could the Yang–Mills model be applied to anything in the real world? It turned out that in this Yang–Mills dream world there are three kinds of photons. One of them is a more or less ordinary photon, and the other two are electrically charged, one positively and one negatively. But actually all three of them are equal: by gauge transformations they turn into each other.

So the question is: do electrically charged particles exist which one could identify as Yang–Mills photons? No way, as Yang and Mills immediately had to admit. The electrically charged photons had to have spin 1, just like the ordinary photon, but their rest masses should vanish. And that does it. Very light, or massless, electrically charged particles would come in copious quantities out of any electricity outlet. Since it costs so little energy to produce such particles, they would be created spontaneously and would try to neutralize any electric field present anywhere.

Electrically charged particles with spin 1 do exist, but they have mass. The best known example is the rho-resonance (ρ) with a mass of 770 MeV. This particle had not yet been discovered in 1954, but its existence was suspected. There are electrically charged ρ^+ and ρ^- resonances and a neutral variety, ρ^0. Could the theory perhaps be altered slightly by adding a few terms to the field equations such that the Yang–Mills photon mass becomes 770 MeV instead of zero? Yang and Mills tried this, but they soon realized that then gauge invariance is lost. This is ugly, and it would become very questionable whether such a mutilated theory would be free of inconsistencies. In short, the Yang–Mills proposal was cute, but was quickly dismissed as being of little direct relevance.

Yet the Yang–Mills article would continue to play an important role. Many investigators realized the importance of such a fundamental idea, that gave such beautiful equations, with the only drawback that nobody knew what, in the real world, they were referring to. For instance, Gell-Mann had clearly been inspired by this theory when he suggested the quark hypothesis, and also the Feynman–Gell-Mann formula for the weak force seemed to point quite clearly to some sort of Yang–Mills principle.

This is what Martinus Veltman was discussing with John Bell in Geneva, in the late 1960s. Why does the weak force seem to be so universal for all particle types? Is it more than a coincidence that the electric charges of all particles are also universal? (Only very few of them have exactly twice this unit of charge.) With a Yang–Mills construction, we would be able to understand all of this much better. After all, the Yang–Mills theory was a direct extension of the theory of electromagnetism. The weak force acts on a charge that seems to obey a conservation rule, also exactly what we have for electromagnetism. Conversely, electric charge only occurs in entire multiples of one universal charge, that of the electron. This is now a feature more characteristic of an extended Yang–Mills system than of a pure theory of electrodynamics alone. It seemed as if what was needed was a unified theory for electromagnetism and the weak force based on the Yang–Mills formalism.

In the mean time, another problem had been solved by a physicist at CERN. The weak force for hadrons was measured always to be a few per cent weaker than that for the leptons. Is this force nearly, but not quite, universal? How can that be? In 1963, Nicola Cabibbo found the origin of this discrepancy.† He calculated that the ordinary hadrons have to share this force with the *strange* hadrons. It is as if this force on the hadrons is very slightly misaligned, and this misalignment is the reason why hadrons with strangeness can also turn into non-strange particles via the weak force, whereas the non-strange hadrons have to settle with the remainder of the force, which is therefore a few per cent less.

Veltman decided that at least *something* of this Yang–Mills theory *had* to be

† The core of his argument had already been suggested a few years earlier by Gell-Mann and Maurice Lévy in a *footnote* of a publication about a particle model with weak interactions in it.

right. If only these charged photons could be given a mass somehow. A generally valid principle is: if a particle that is responsible for transmitting a force ('force carrier') is given a certain amount of mass, then the force transmitted by this particle becomes one of limited range. The larger the mass, the shorter the range. According to the theory for the weak force, that of Gell-Mann and Feynman, a weak current generates instantly and on the same spot another weak current, as if a force were active with an extremely short range. It was generally known that a very *heavy* version of a photon could transmit just such a force. This was the 'intermediate vector boson theory' for the weak force. It would have to be a very heavy, electrically charged particle with spin 1, the so-called W^+ particle and its antiparticle, the W^-.

Every weak interaction process could then be seen as resulting from two successive interactions. First a particle makes a transition to a fellow member of the multiplet it is in (for instance, a neutron changes into a proton), at which point a W particle is produced. As the mass of this W is far too big, and the energy set free at the transition is usually much too small, for this W to be released, it can exist only during a tiny fraction of time, but during this time it can undergo a second interaction. Either it can be absorbed by another particle that will then undergo a similar transition to a member of its multiplet, or else it can decay into a particle and an antiparticle, for instance an electron (e) and an antineutrino (v_e); see Figure 11(b).

What happens now if we take the Yang–Mills equations and change them just enough to give the charged photons a mass? Gauge invariance is then *approximately* right. We have seen such a thing elsewhere in Nature. There are all sorts of symmetries that are only approximately right. We had, for instance, isospin symmetry, according to which protons and neutrons can be treated equally; yet protons and neutrons are not *exactly* equal.

The Yang–Mills equations that Veltman decided to investigate had two kinds of terms in them, those which were gauge invariant and only one term which was not. The system resulting from this had already been studied – Richard Feynman had tried it. But Feynman's motivation had been a quite different one.

To see what he did, we must go back to 1961. As was already clear at that time, one of the most difficult problems in theoretical particle physics, perhaps *the* most difficult problem, is to reconcile the rules of quantum mechanics with Einstein's general relativity. Feynman wanted to give it a try. Would he have done this had he known that, even today, the problem is unsolved? Anyway, he quickly discovered that even the simplest calculations in gravity theory gave rise to long, immensely complicated expressions. This time it was Gell-Mann who had a useful suggestion for Feynman: why did he not try his skills first on Yang–Mills theory, this being technically much easier, whereas its most crucial property, gauge invariance, looks very much like the coordinate invariance of gravity theory? Just like Yang–Mills theory, gravity has its gauge bosons, gravitons. Both of these 'force carrying particles', the graviton and the bosons of the Yang–Mills theory, are massless, moving about with the speed of light.

And thus Feynman turned his attention to Yang–Mills theory. That the gauge bosons were massless was a nuisance to Feynman. He saw no harm in giving these particles just a 'tiny' mass. Feynman unscrupulously added a tiny mass term to the equations for his particles – the same term Veltman would consider later – and he started to calculate. Feynman was a true expert in simplifying complicated calculations and arguments. He discovered that you could simplify the rules if you added what he called 'ghost particles'. If particles collide, the situation at the end, the 'out-state', is a result of many intermediate interactions. If you sweep the various terms together in a special way, it just looks as if there are extra particles being temporarily produced and successively annihilated. But these particles are ghosts. What remains after the collision are only those particles that the experimenter can observe: the 'physical' particles.

Feynman did not finish his work. He found that, after the first series of correction terms, the next series of corrections caused things to become really complicated. The only publication of his work was a set of lectures Feynman gave in Poland, in 1962. Notes were taken by a student, and these appeared in the scientific journal *Acta Physica Polonica*. It was not easy to figure out from these notes how Feynman had really obtained his results.

Among those present at his Polish lectures was Bryce DeWitt (an American obviously with Dutch ancestors). DeWitt was not very interested in the Yang–Mills problem but in quantum gravity. It is interesting to read in the proceedings how DeWitt tried to persuade Feynman to disclose the details of his calculations. In Feynman's opinion it made little sense to write long and complicated formulae on the blackboard if nobody could follow them anyway, but DeWitt insisted. Feynman finally gave in. 'I can get as much unintelligible gibberish on the blackboard as anybody else,' he said, and then came the truly interesting details.

From what happened next it turned out that DeWitt had understood very well what Feynman had done. He carried on where Feynman had become stuck. He improved the techniques and found the correct calculational rules for the successive corrections at all orders. In 1964, he published only the final results, but these attracted little attention because it was difficult at the time to understand what they said. Also, as would become clear later, the results were not totally correct. In 1967, DeWitt published his complete derivation, but this became three articles, each being so lengthy and detailed that everyone lost courage even before starting to read them. Only later did we come to understand how he had derived the correct rules; there were even techniques in those papers that were remarkably close to modern methods introduced much later by others.

In the mean time others had become interested in the gravity/Yang–Mills problem. Stanley Mandelstam in California had his own, very different, approach. He claimed to have a new formalism for gravity and for Yang–Mills theory, from which the Feynman rules, including Feynman's ghost particles, followed. Also there were groups in the USSR. In Moscow, several scientists were making progress; in Leningrad, Ludwig D. Faddeev and Victor N. Popov came up with practically the same rules as Mandelstam. Their arguments looked very formal (and therefore suspect, in Veltman's eyes) but their first publication was short, crystal clear and easy to understand.

We could work with this. But not everything was in order as yet. Mandelstam's rules were not exactly the same as those of the Russians. And all of them had a factor of 2 different from Feynman. 'Who cares about a factor two', Feynman would say about this later (it was not in his nature to be much bothered about such petty details). But now we know that this difference came about because Feynman's theory, with its mass term, was fundamentally different from those of DeWitt, Mandelstam and the Russians, and that had also been the reason why Feynman had been unable to proceed.

It also became clear that there is a very fundamental difference between gravity theory and Yang–Mills theory. The rules for Yang–Mills theory were obviously suitable for a renormalization program, whereas this would never work for gravity. In gravity, an infinite series of infinite counter-forces would be needed, and as long as we do not understand the physical nature of these forces, we will not be able to produce a powerful theory. Therefore we will give up on gravity theory for the time being, and return to the Yang–Mills system.

What Veltman wanted was a Yang–Mills theory with a mass term in it, but in such a way that it would be renormalizable. This was exactly the model Feynman had started off with without complete success, whereas the models of the others had been strictly massless. Veltman knew that only the first correction terms could be implemented using Feynman's procedure. These terms suggested that this theory should indeed be renormalizable. After all, at first sight, the mass term Feynman had added looked rather innocent. But what the others had done seemed to be inapplicable here. It seemed that Veltman had fallen into an ugly mess. The equations were not difficult, just very long. It had to be possible

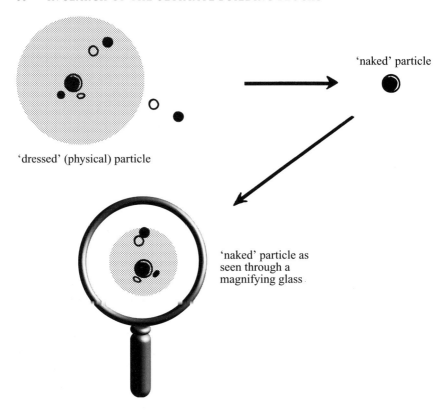

'naked' particle

'dressed' (physical) particle

'naked' particle as
seen through a
magnifying glass

Figure 9.

somehow to do the demanding yet boring manipulations by computer. Indeed
the higher order corrections should follow from the theory unambiguously. All
one had to do was apply the 'dispersion relations' that I mentioned earlier in
Chapter 9.

Veltman carried out pioneering work in designing computer programs for
algebraic calculations. When, finally, the computer programs worked flawlessly,
there was a disappointment: the infinite forces did not cancel out. In other words,
the 'massive Yang–Mills theory' is not renormalizable, and is therefore useless as
a theory for the weak force.

Why are some theories for elementary particles renormalizable and others
not? To clarify the situation, recall Chapter 1. There we remarked that we can
view our world through a magnifying glass, and then the world of the small
seems to be a replica of the world of large things. This situation also occurs in

the world of elementary particles. At a small scale, the particles behave nearly the same way as at a larger scale (although we have to keep all sizes below the size of an atomic nucleus). Electromagnetic forces, for instance, continue to obey the same equations. And what, at larger scale, can be seen as a 'naked' particle, when viewed under a 'microscope' will turn out to be surrounded by other particles like photons after all. To see a particle that is also naked under a microscope, you would have to undress it further. This is the way the philosophy of renormalization works: the concept of a 'naked' particle is merely a relative one. In practice, a theory turns out to be renormalizable if a dressed particle follows the same laws as a naked particle when seen under the microscope.

Only the mass of a particle turns out not to remain the same when you look at it through a microscope (or, in the jargon of theoretical physics, perform a scale transformation). This is because the *range* of a force seems to be larger under a microscope; correspondingly, the mass of a particle seems to be smaller. Note that this situation is the reverse of that of ordinary life, where a sand grain seems to be larger – hence heavier? – when viewed through a microscope.

A consequence of all this is that, in a Yang–Mills theory, the mass term will seem to disappear when you perform a scale transformation. This implies that under a microscope gauge invariance is restored. This is the factual source of the difficulty Veltman was confronted with. Is the Yang–Mills vector potential directly observable or not? It seems to be observable in the world of large things but not in the world of the small. This is a contradiction, and provides the reason why the scheme could never have worked properly.

11 Superconducting empty space: the Higgs–Kibble machine

There was a way out! However, it originates in quite a different branch of theoretical physics, to wit the physics of metals at very low temperatures. At very low temperatures, 'quantum phenomena' can lead to quite remarkable effects. To describe these effects one uses quantized field theories, exactly the same as the ones employed in elementary particle physics. The *physics* of elementary particles has *nothing* to do with the physics of low temperatures, but the *mathematics* is very very similar!

The 'field' in some material that becomes important at low temperatures could be the field describing how atoms there oscillate about their equilibrium points, or else the field that describes electrons in this material. At very low temperatures we have to deal with the 'quanta' of these fields. For instance, the 'phonon' is the quantum of sound. Its behavior resembles that of the photon, the quantum of light, except that the numbers are very different: phonons travel with the speed of sound, hundreds or perhaps thousands of meters per second; the photon travels with light velocity, which is 300 000 *kilo*meters/second, nearly one million times faster! The elementary particles we are interested in usually have velocities close to that of light.

One of the more spectacular 'quantum phenomena' that can take place in strongly cooled metals is called *superconductivity*, which is the fact that the resistance of such a material to electrical currents vanishes completely. One consequence of this is that the material does not tolerate even the slightest electric potential difference, because it would be neutralized immediately by an 'ideal' electrical current. In addition, this material also does not tolerate the presence of magnetic fields, because, according to Maxwell's equations, switching on a magnetic field would be associated with induction currents. As long as these currents meet no resistance they will completely neutralize the magnetic field. Because of this, no electric or magnetic field can be created inside a superconductor. This situation only changes if the induced currents become too strong, which happens if the superconductor is exposed to the fields of *very strong* magnets; in this case the superconductor becomes disturbed. Not being able to stand up to brute force, it will lose its superconductivity, and it will surrender by allowing the magnetic field in.

So what does a superconductor have to do with elementary particles? Well, superconducting material could be viewed as a system in which the electromagnetic field is *a field with a very short range*. It is being screened. And yet it is Maxwell's field, a gauge field. This is what makes the superconductor interesting for someone wanting to describe the weak force among particles as a gauge theory! What a nice feature of theoretical physics! Totally different worlds can be compared with each other, just because they happen to obey similar mathematical equations.

How does the superconductor work? The true cause of this peculiar phenomenon was disentangled by John Bardeen, Leon N. Cooper and John R. Schrieffer (who received the Nobel Prize for their work in 1972). Two special conditions have to be realized by the electrons in a solid piece of material which, when taken together, can give rise to superconductivity: the first is *pairing*, and the other is *Bose condensation*.

'Pairing' implies that electrons form pairs and act pair-wise, with the phonons producing the force that keeps the pairs together. The two electrons in a pair each spin about their own axis, in opposite directions, such that the pair (called a 'Cooper pair') as a whole no longer shows any rotation ('angular momentum'). A Cooper pair behaves as a 'particle'† with spin 0 and electric charge −2.

'Bose condensation' is a typical quantum mechanical phenomenon. It can only apply to particles of integer spin (bosons). Like lemmings, bosons will all plunge together into the lowest energy state possible. Remember: bosons like to all do the same thing. In this state they can still move, but no further energy can be taken from them. A consequence is that there is no resistance to this motion. The Cooper pairs move freely, so that electric currents may exist. These currents do not meet with any resistance. A similar phenomenon takes place in liquid helium at very low temperatures. Here it is the helium atoms themselves that

---

† Remarkably, the electrons in such a pair are only very loosely bound together. The situation resembles a dance floor where very wild pop music is played: it is hardly possible to recognize who belongs to whom in a pair. Schrieffer later explained why his theory was so difficult: you have to write the choreography for a dance with more than one million times one million times one million dancing pairs.

Bose condense. The consequence is that this liquid can stream through the tiniest holes without the least amount of resistance.†

Because separate electrons have spin $\frac{1}{2}$, they themselves cannot Bose condense. Particles whose spin is an integer plus one-half (fermions) must all be in *different* quantum states, because of Pauli's exclusion principle. This is why superconductivity can only occur if pair formation has taken place. Yes, I realize that this will raise a few questions, and I apologize beforehand. I have again tried to translate formulae into words, which implies that the reasonings may sound very unsatisfactory. Just look at this as a somewhat unwieldy 'quantum logic'!

That superconductivity could be of importance for elementary particles was discovered by the Belgian François Englert, the American Robert Brout and the Englishman Peter Higgs. They proposed a model for elementary particles in which electrically charged particles without spin undergo Bose condensation. This time, however, the condensation takes place not inside some material, but in empty space (the 'vacuum') itself. The forces among these particles have then been chosen in such a special way that it saves energy to fill the vacuum with particles rather than keeping it empty. These particles are not directly observable. We would experience this state, in which space and time are sizzling with Higgs particles (as they are now called), but in which the energy is as low as it can ever be, as if space-time were completely empty.

The Higgs particles are the quanta of the 'Higgs field'. A characteristic of the Higgs field is that the energy in it is lowest when the field has a certain strength, and not when it is zero. What we experience as empty space is nothing but the field configuration that has the lowest possible energy. If we move from field jargon to particle jargon, this means that empty space is actually filled with Higgs particles. They have 'Bose condensed'.

This empty space has lots of properties in common with the interior of a superconductor. The electromagnetic field here also has a short range. This is directly related to the fact that in such a world the photon has a certain amount of rest mass.

And yet we have a complete gauge symmetry; gauge invariance is not violated anywhere. And thus we have learned how to turn a photon into a 'massive' particle *without* violating gauge invariance. All we had to do was to add these Higgs particles into our equations. The reason why the effect of gauge invariance on the properties of the photon is so different now is that the equations are

---

† But there is a restriction on the *speed* of the flowing atoms, which must stay below a certain limit. Only if the speed is below this limit is the resistance zero.

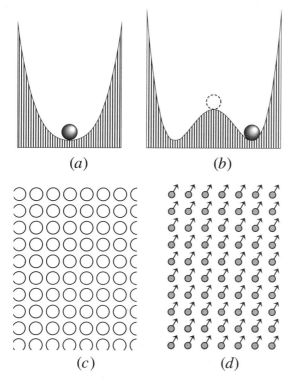

Figure 10. Symmetry and spontaneous symmetry breaking. A symmetric system of equations may have a symmetric solution which is also stable (*a*), or else the system would end up in an asymmetric solution (*b*). If a *field* has a stable symmetric solution, the world it describes will be symmetric (*c*). If the stable solution is asymmetric, all phenomena in this world will also show this asymmetry (*d*).

completely altered by the presence of the Higgs field in our vacuum state. Sometimes it is said that 'the vacuum state breaks the symmetry spontaneously'. This is actually not correct, but the phenomenon is closely related to spontaneous symmetry breaking in other situations.

Now Higgs considered only 'ordinary' electromagnetic fields. But of course we know that the ordinary photon in a true vacuum does *not* have a rest mass. It was Thomas Kibble who proposed to make a Yang–Mills theory superconducting in this way, just by adding a spinless particle with Yang–Mills charge instead of ordinary charge, and to assume that this particle Bose condenses. Then the Yang–Mills forces are reduced to act over a short range, and the Yang–Mills photons become massive spin 1 particles.

But was this not the ideal solution to the problem of the previous chapter? The Yang–Mills photons get their masses and the gauge principle remains valid! I think there were two reasons why this approach at first did not receive the attention it deserved. First of all, people thought the scheme was ugly. The gauge principle was still there, but it was no longer the central theme. The Higgs field had been put there 'by hand', and the Higgs particle itself was not a 'gauge particle'. If that is allowed, why not introduce a host of arbitrary particles and fields? These ideas were viewed as simple toy models without much fundamental significance.

Secondly, there was what was called 'the Goldstone theorem'. Particle models with 'spontaneous symmetry breakdown' had been proposed before, but for a large class of such models Jeoffrey Goldstone had proved that they always contain *massless particles without spin*. Many researchers therefore thought that the Higgs theory ought to also contain such a massless Goldstone particle, and this would be a nuisance because there are no Goldstone particles among any of the particles we know.† Now, even Goldstone himself had warned that the Higgs model did not satisfy the conditions for his proof, so that it need not be valid for that case, but everyone was so impressed by this solid piece of mathematics that the Higgs–Kibble model was not popular for quite some time.

And so the Goldstone theorem was used as a 'no-go theorem': if empty space is not symmetric, then you cannot avoid the presence of massless, spinless particles. Now we know that, in our case, the small-print disables the theorem; the Goldstone particles become invisible due to gauge invariance, they are nothing but the 'ghost particles' Feynman first encountered in his calculations. Besides, remember that I stated earlier that the Higgs mechanism is not truly a spontaneous symmetry breaking.

That one surely can construct realistic particle models in which the Yang–Mills system is responsible for the weak force, and in which the Higgs–Kibble mechanism is responsible for its short range, had been suggested independently by two prominent investigators. One of them was the Pakistani Abdus Salam. Salam was looking for esthetic models for particles, and he thought that the Yang–Mills idea was so beautiful that this was sufficient reason to try to build it into a model for the weak interactions. The carrier of the weak force *had* to be a Yang–Mills photon, and for him the Higgs–Kibble mechanism was the only acceptable explanation for the fact that this carrier had a certain amount of rest mass.

† *Nearly* Goldstone particles do exist; see the next chapter!

In a meeting sponsored by the Swedish Nobel consortium in 1968, Salam explained the ideas he had been hatching together with his co-author John Ward. His lecture, and the ensuing discussion, were published, and the question that soon became the center of the discussion was whether this theory was renormalizable. Intuitively, Salam understood that the answer had to be affirmative, but he could not provide any detailed proof of this. He was unable to formulate the Feynman rules, and had to admit that if the theory is taken at face value it seemed to be full of ghost particles that were about to spoil everything. If the production rate of such particles in some experiment is computed, either a 'negative rate' results, or the energy turns out to be negative. Both would be unacceptable if such a theory is to be logically coherent.

The other investigator who had reached more or less the same point was the American Steven Weinberg. But Weinberg went one important step further. He formulated in much more detail a simple model in which he indicated accurately which fields existed and how they would interact. But he limited himself to the leptons. Weinberg realized that besides the ordinary photon there had to be *three* heavy Yang–Mills photons: one positively charged, one negatively charged and one neutral. As for the charged photons, everybody agreed that these would be needed to describe the weak interactions; they would be the famous intermediate vector bosons, W^+ and W^-. According to Weinberg their masses had to be more than $60\,000\,\text{MeV}$. But these charged vector bosons alone would suffice to explain *all* weak interaction processes known at that time. That apart from these, and the ordinary photon, γ, you actually also needed another neutral component (Weinberg called it Z^0), was not at all evident. It was found that the Z^0 mass had to be a little more than that of the charged bosons. Figure 11 indicates how it was suggested the weak interactions should take place. However, it was well known that 'neutral' exchange processes were never observed, so one had to conclude that diagram of Figure 11(c) should be forbidden for some reason or other!

This was a problem for those who wanted to believe in a neutral Z^0 particle. This problem becomes even more striking if the π^- lifetime is compared with the K_{Long} lifetime. Why does the pion decay into a μ^- and a ν_μ, whereas K_{Long}, during its much longer life, never decays, via a Z^0, into for instance a μ^+ and a μ^-? Weinberg saw, however, that the strict mathematical rules of the Yang–Mills system surely demanded the existence of a current that can emit neutral Z^0 particles. He assumed that apparently something was not yet in order with the hadrons. This is why he entitled his publication: 'A Model for Leptons'. For the leptons the most important new consequence of the existence of the neutral Z^0 particle was the collision process $\nu_\mu + e^- \rightarrow \nu_\mu + e^-$ (Figure 12).

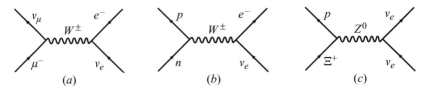

Figure 11. Diagrams showing how a weak interaction can be brought about by exchange of an intermediate vector boson W^-. Part (a) shows that the transition $\mu^- + v_e \rightarrow v_\mu + e^-$ can take place either via the intermediate state $v_\mu + W^- + v_e$ or via $\mu^- + e^- + W^+$. If we follow the arrows in the other direction, we see the interactions of the corresponding antiparticles. The diagram then shows how μ^- can decay into $v_\mu + e^- + \bar{v}_e$ (the antineutrino). Part (b) shows the decay $n \rightarrow p + e^- + \bar{v}_e$, and part (c) shows how a neutral force carrier could generate a decay of Σ^+. This particular decay process is not observed however!

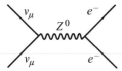

Figure 12. Muonic neutrinos can only collide (elastically) against electrons if a neutral particle Z^0 is exchanged.

Weinberg concluded therefore that his theory could be checked by experiment. Although neutrino experiments along these lines had already been performed, the existence of interactions of this sort was still very uncertain. Actually the efficiency with which electron-type neutrinos collide with electrons would also be affected by the contribution due to Z^0 exchange, but this process would also take place by charged exchanges.

Weinberg also supposed that his model would be renormalizable, but he could not formulate the detailed mathematical rules. This was in 1967. By 1970, both Weinberg and Salam had lost interest in the Yang–Mills theory. New theories for the weak interactions had arrived, in which diagrams other than the ones of Figure 11 would play dominant roles; theories in which there were infinitely many diagrams, and theories in which negative probabilities or a slight violation of causality would be allowed. Now, with hindsight, it is easy to say why such ideas were doomed to failure, but at that time all possibilities and impossibilities had to be checked.

This was more than enough work for a young investigator such as myself. In comparison to others I read little and thought a lot. In this way I ran the risk of thinking very long and very deeply only to discover something that turns out to be already long known, but it did give me the advantage that I understood the basic problems inside out. This is how I fared with the Higgs–Kibble mechanism (I do not think I knew at that time that this is what it was called). Veltman was very skeptical about such ideas; it was not easy to convince him that what we call empty space is actually filled with invisible particles. Would these, he said, not betray their presence by their gravitational fields? The theory *can* be formulated in such a way that these gravitational fields are exactly compensated for by other invisible particles or by a mysterious contribution from empty space itself. Exactly how Nature does manage to mask these gravity effects so efficiently and completely that we fail to notice anything at all is a mystery that continues to be hotly disputed to this day. In my opinion, the resolution of this riddle must be postponed until we understand much better the theory of quantum gravity itself. And that has not happened yet.

12 Models

So I liked the Higgs mechanism. But is it correct? Does it yield a healthy theory? What is and what is not allowed in describing elementary particles, and why?

Various universities in different countries organize courses at an international level, called 'summer schools', where in prestigious, yet quiet, recreational resorts experts and students convene to give lectures, and listen to them, and most of all to discuss their research subjects. The most prestigious summer school in my subject was in Les Houches, a winter sport resort near Chamonix, on the slopes of Mont Blanc. Presumably because I was too late in registering there, I was not admitted. But I was admitted at my second choice: Cargèse. I went there in 1970.

The Institut des Sciences near Cargèse on the island of Corsica was established by the French physicist Maurice Lévy. On a beautiful piece of land with a little beach, a small building had been placed where, from 1960 onwards, summer schools and conferences had been held. The 1970 summer school would be about the *strong* interactions. Together with Gell-Mann, Lévy had constructed a model for the strong interaction. Although it was not expected to represent the complete truth, it had the nice feature that all symmetries of the strong interactions were reproduced in a very interesting way. But, because the forces are so strong, the particles will not even approximately move in straight lines, and that made the usual approximation scheme, called 'perturbation expansion', unreliable. The discussions would be dominated by attempts to extract from the model results that were nevertheless meaningful, but some fundamental aspects of the model, with its endless series of approximations not converging to anything, would also be considered.

And the model was interesting. It was a renormalizable model in which protons, neutrons and the three pion types would play the leading roles. But a fourth companion of the pions was needed which was called 'sigma' (σ). The symmetry of the model required that the 'naked' protons and neutrons be *massless*. Only then would one be able to understand how the currents work on which the *weak* force acts. And here is the interesting bit: it was assumed that the sigma particles undergo Bose condensation. So here also we had 'spontaneous symmetry breaking'. The protons and neutrons, which would be massless in a symmetrical surrounding, were now slowed down, as it were, by all those sigma particles populating the vacuum; they get a mass because of this spontaneous symmetry breakdown. For this system Goldstone's theorem was true: the three pions become Goldstone particles, and therefore they lose their rest masses.

Now this was not so bad at all. It had already been noticed that of all the hadrons the pions are the lightest by far. In most theories it was the *squares* of the masses of the particles that had to be compared. The square of the pion mass is nearly fourteen times smaller than the mass of the next particle, the kaon. An approximation in which the pion mass is put equal to zero is therefore not so crazy. In an improved version of the sigma model, we may assume the presence of a tiny disturbance that gives the pion its mass. Later, when we will make use of the quark model, both the Goldstone pions and the sigma particle will each be considered as consisting of a quark and an antiquark. The 'disturbance' that gives a mass to the pion then turns out to be the tiny mass of the u and the d quarks inside a hadron, as is indicated in Table 7 on page 115. This interpretation was already well understood in 1970.

Two authorities on the subject of renormalization, the Korean Benjamin Lee and the German Kurt Symanzik, came to Cargèse to explain how the renormalization procedure had to be carried out in the sigma model, such that its most important property, spontaneous symmetry breaking, would not be spoilt.

'Now can one do the same thing if there is a Yang–Mills field?', I asked both Lee and Symanzik. Both gave me the same answer: if I were a student of Veltman's I should ask *him*; they had not studied Yang–Mills fields. But it dawned on me that the prescriptions as formulated by Lee and Symanzik for the sigma model should remain applicable for any system in which symmetry is affected by Bose condensation, including the Higgs–Kibble theory. Here I had the seedling of a new theory. But how could I make it grow?

Veltman knew exactly how the requirements for a renormalized theory for massive Yang–Mills particles should be formulated. The theory had to give an unambiguous prescription enabling one to calculate the probabilities of a certain initial state evolving into a certain final state, and these probabilities must all add up to exactly unity. As soon as the slightest mistake is made in the formalism, this test will fail. Also, every particle configuration must have an energy that is greater than zero, otherwise every configuration will be unstable. This latter statement implied that it does not make one bit of difference if 'empty space' is filled to the brim with invisible particles or not. The total energy is the lowest possible, and that is all that counts.

When I returned from Cargèse to Utrecht, I knew what I wanted to investigate and what my thesis would be about. First, the precise prescriptions should be worked out as to how 'pure' Yang–Mills theory, without any Higgs–Kibble mechanisms, had to be renormalized. True, Feynman, DeWitt, Mandelstam, Faddeev, Popov, and others had given the Feynman rules, but they had not explained whether and how these obeyed Veltman's requirements. Furthermore, there were small differences in the various rules. Were they all talking about the same theory? Were the prescriptions unique? Could one calculate the same thing in different ways?

This was the subject of my first publication, which emerged after extensive discussions with Veltman. Indeed, it turned out that the rules of Mandelstam on the one hand and Faddeev and Popov on the other do describe the same theory, and that the renormalization procedure could be carried out, although some bits and pieces of a complete proof were still missing. The theory was not at all yet as perfect as quantum electrodynamics and what Lee and Symanzik had made out of the sigma model, but the beginning was there. And now I also knew how to include the mass.

It was clear to me now how to put the mass in. For all proofs of the massless case, gauge invariance had been absolutely essential. If the mass was included in the way Veltman wanted to do it, gauge invariance would be disturbed, and nothing would work anymore. Absolutely the only way was via the Higgs–Kibble mechanism. And that was to become my second publication, the observation that a theory with a Higgs–Kibble mechanism is still renormalizable and that the delicate techniques Veltman had constructed could be applied there. The first application I thought of was a version of the Gell-Mann–Lévy model that now also contained a rho meson. This after all was the kind of particle that Yang and Mills themselves had wanted to describe (see Chapter 10), but at the time they did not know where its mass came from. Also, several people in Cargèse had talked about their hopes of having a vector particle such as the rho in their models. In this way, one might obtain improved renormalizable field theories for the strong interactions.

But even if such models for the strong interactions were useful, they would not present a breakthrough. The rho particle is also a meson containing a quark and an antiquark; this does not fit very well with the picture of the rho as a gauge particle. It behaves a little bit like a gauge particle, but its intestines are different.

No, Veltman convinced me that the *weak interaction* was much more important.

This meant taking exactly the same equations, but now letting them stand as a model for entirely different particles in physics, the W and the Z particles, each having a mass more than one hundred times larger than that of the rho particle.

In fact, we now had the most essential part of the proof that the model Weinberg had written down in 1967 was renormalizable.† For Veltman it was important that things could also be rephrased in such a way that one never spoke of invisible particles populating empty space; Veltman was still worried about the gravitational fields of those invisible particles, so I transformed them away from the equations. The new rules could be fed into his computer program, where it soon became clear that everything worked. The resolution of the tough problems had been found!

A good stroke of fortune had it that, in January 1971, a large International Conference on Particle Physics would take place in Amsterdam. The weak interaction models that would be discussed there were not all that inspiring. It was to Veltman's great delight that he could come forward with something new up his sleeve, giving me the opportunity to announce the fresh proofs that the Higgs–Kibble models were renormalizable. A period of very intense collaboration with Veltman followed. My 'proof' was not yet totally convincing. Together we would complete, polish and generalize the techniques we had found. We were also able to make progress in the area of quantum gravity, albeit modestly.

Later I would look back to 1970 as the year in which I made two big discoveries that would profoundly influence the rest of my life. One was the proof that I have just described, the formulation of which led to my appointment as an assistant professor at the University of Utrecht. The second was the girl who soon became my wife. During the years that followed, I was invited to numerous places to lecture about gauge theories, among others in Cargèse and Les Houches, and I had two daughters.

Gauge theories quickly gained in popularity. Benjamin Lee, Steven Weinberg and many others became enthusiastic adherents. Kurt Symanzik was the first to invite me abroad to give a lecture about gauge theories, in Hamburg.‡

Now there was a lot to be done. First of all, which model of the weak force should we actually believe? The point is that the weak interaction theory that

---

† Apart from a technical detail, the so-called 'anomalies'. In modern versions of this model, this point is also in order. See Chapter 17.

‡ I adored Symanzik, and this visit to Hamburg was a memorable one. Symanzik had all the walls of his appartment – or at least those that were not hidden behind scientific books and journals – covered with posters, mostly of attractive women. The largest two posters were 'his two heroes': Albert Einstein and Brigitte Bardot.

Weinberg had proposed, and which I described in the previous chapter, was merely the most obvious choice, but not at all the *only* possibility. And those quarks were not yet quite in order. It was possible to think of other varieties of the same scheme that would also be renormalizable. The most important thing that had been achieved was that now the *general rules of the game* for constructing a model, *any* model, or theory, had become clear. The fundamental conditions for renormalizability of a particle theory could be formulated as follows:

- There are fundamental particles with spin 1, which we could call 'photons'. These *have* to be of the Yang–Mills type. They are the main carriers of forces over larger distances. All other particles, and indeed also the photons themselves, feel these forces because they carry 'charges'. The precise mathematical formulation makes use of 'group theory', which would be too complicated to explain any further here. It is important that only a few constants of Nature are required to be able to calculate all interactions. The values these constants of Nature have cannot be derived from the theory, but have to be determined from experimental data.

- There are fundamental particles with spin $\frac{1}{2}$. We call them 'fermions', or, rather, 'Dirac fermions'. Two types would be introduced: the leptons and the quarks, but in principle there could be as many types as is wanted. Depending on how they interact with the Yang–Mills fields, they may or may not have a 'naked' rest mass. These masses then also have to be determined from experiment; they cannot be derived from the theory.

- There are fundamental particles with spin 0. Their interactions with the Yang–Mills photons are determined by their charges, but new constants of Nature are required to describe their mutual interactions and their interactions with the fermions. If we allow for many spin 0 particles, there would be very many constants of Nature whose values would, in principle, be unpredictable. Usually we assume the existence of only one or two types of spin 0 particles, but also their number can *a priori* be anything. Essential now is that these particles can undergo Bose condensation so that we get a Higgs–Kibble mechanism. By this Bose condensation, most of the Yang–Mills photons, as well as the originally massless fermions, are assigned definite, finite values for their rest masses. Since these values are always related to constants of

Nature that are, in principle, arbitrary, we will, in most cases, not be able to calculate these masses. They have to be determined (directly or indirectly) from experimental observations.

- There is a technical constraint: the 'anomalies'. Not all combinations of fermions are admissible. I will say some more about this in Chapter 17.

- Fundamental particles with higher spin are not allowed. There can be 'bound states', consisting of several fundamental particles orbiting around each other, and such systems can rotate much faster than the fundamental particles themselves. Only in this way can there exist particles with spin $1\frac{1}{2}$, 2, $2\frac{1}{2}$ and so on. (An exception would be the graviton, having spin 2, and perhaps the gravitino with spin $1\frac{1}{2}$, but these theories cannot (yet?) be formulated precisely, because they are not renormalizable.)

So, we can see that the amount of freedom and arbitrariness is rather large. How many fundamental particles and Yang–Mills fields are there? And how do we determine from today's experimental observations which combination is the correct one? A period began which I like to characterize as 'The Great Model Rush', a race for models. The first author to come forward with the correct model could surely count on a Nobel prize, and now suddenly the rules for constructing all possible models were known. The craziest ideas were suggested. Curiously our Great Creator† did not have this much imagination.‡ Weinberg's original version, in some sense the simplest possible structure, turned out to be correct as far as the leptons are concerned. But what about the difficulty Weinberg had? Why does the neutral component of the Yang–Mills photons not cause particles such as Σ^+ to decay into a proton and two neutrinos? Well, the answer to this turned out to be the one already given by Glashow, Iliopoulos and Maiani in their 1969 article. Introduce a fourth quark, named 'charm', and then the unwanted interactions will be washed away. The GIM mechanism was directly applicable. But now it could be calculated how efficiently this cancellation takes place, and it was found that it only works to the required precision if the charmed quark is only slightly heavier than the quarks already known. K_{Long}, in particular, is 'sensitive' to this. If the charmed quark were very heavy, we should see many more decays

---

† This wording I took over from my high school physics teacher. I apologize: no conclusions are to be drawn from this concerning my religious beliefs.

‡ You would not perhaps say that when you look at what He did when He was creating humanity. But perhaps what He used here was His infinite sense of humour.

of K_{Long} into two muons. In reality only one out of one hundred million is seen to decay this way! This would prove to be an important observation. It was Glashow in particular who emphasized that experimenters should search for particles containing the new quark, and he tried to indicate which phenomena the experimenters should be looking for.

So, although by introducing charm the unwanted effects of Z^0 could be eliminated, one could still expect new phenomena that would be brought about by the Z^0. In particular, neutrinos would experience forces because of Z^0 exchange. If a neutrino captured a Z^0, it would continue in a different direction, but it would remain a neutrino. We call this an 'elastic' collision. If there were only charged W particles, the neutrino could only undergo interactions in such a way that it changes into an electrically charged lepton.

Experimenters happened to have machines ready to detect elastic neutrino collisions. Actually they had already looked for such events, but with a negative result. Did this imply that those collision processes, which were being called 'neutral current processes', did not exist? Well, what had been investigated were elastic collisions of muonic neutrinos against atomic nuclei. This is a terribly difficult experiment, because these same neutrinos can also collide against atomic nuclei such that they *do* turn into a muon, and this muon might escape detection. Furthermore they could produce a *neutron*, which in turn could mimic an interaction a little further on that would be hard to distinguish from the *neutrino* interaction being searched for. In an experiment to detect elastic neutrino collisions, it is very difficult to determine whether the phenomena observed as a result of the experiment are actually what is being sought, or whether they just look that way.

All this was very well known to the experimenters. The difficulty lay in estimating how strong these false effects and other disturbing sources of 'background noise' were, and then in subtracting them from those events that were registered, so as to determine how strong the real 'signal' was. In 1972 the widely held opinion among experimenters was that the signal was 'probably' not there.

But now that the theoreticians were suddenly so keen to have more certainty about this, and strongly hinted that there should be an effect, it was admitted that there had been essentially only one experiment, and that its conclusion had not been very convincing. Fortunately, there was a good opportunity to do the experiment again, this time with much greater accuracy, at CERN, where a new, gigantic particle detector had been completed. In a container filled with freon and surrounded by thousands of tons of steel and copper, tracks consisting of little bubbles caused by fast moving particles could be made visible and photographed.

The machine was called 'Gargamelle', after the mother of the giant Gargantua in Rabelais' *Gargantua et Pantagruel*.

Paul Musset was one of the researchers who began the time-consuming analysis of the thousands of photographs. The first announcements of the results were shrouded in uncertainty. The first claim of success was withdrawn, so that critics sarcastically remarked that the neutral currents were going on and off. A neutral alternating current perhaps? In 1974, however, after a very thorough analysis, the existence of neutral current events could be confirmed with practical certainty.

Direct elastic collisions between neutrinos and electrons are extremely rare. Analysis of CERN data by Hellmut Faissner in Aachen nevertheless yielded convincing pictures of electrons that were suddenly torn away from their atoms by a neutrino.

Had the theory now genuinely been confirmed, or did experimenters surrender under pressure from the theoreticians? Were these results real, or merely a product of wishful thinking? The latter is sometimes suspected by critics and historians of science. It seems that if theory doesn't want it, it isn't there, and if theory requires it experimenters suddenly all see it.

I want to say two things here. First, to suspect that they are so very prejudiced is a severe accusation against any respectable experimenter. A good research scientist will rather underestimate than overestimate the accuracy of his results, most particularly if the observation is of this much importance. Of course, regrettably, it happens every now and again that someone overestimates his precision. In most cases his colleagues will be quick to point this out to him. Even if theory 'wants' the result.

Secondly, the experimental results were indeed correct. Other observations would later confirm, in several different ways, that everything was in order down to the finest details. There are philosophers of science who claim that every scientific discovery will sooner or later be overturned by 'scientific revolutions', and that therefore there is no such thing as an 'absolute truth'. One of them also argued that the question of whether or not neutral currents exist is solely a question of 'this moment's opinion among scientists', and does not have to be a truth that can be established for ever. And, yes, I presume that the once popular flat Earth theory, claiming that all pictures taken from spacecrafts are fakes, will also pop up again some time in the future. In fact, the existence of neutral current events can be established no less objectively than the fact that the Earth is round.

So gradually the situation concerning the magnificent theory for the weak force became increasingly clarified. Weinberg's model had persevered in the light of the new experiments. That the prediction about neutral currents had come true had encouraged the theoreticians. But in quark theory we needed another quark, 'charm', and no single particle had been detected that could contain such a charm quark. And, for the GIM mechanism to function properly, we really needed a charm quark that was not too heavy. This implied that it should be possible in due time to produce and study charmed particles in the laboratory. This now was the new prediction, and it would indeed soon come true. But something else happened: a breakthrough in the understanding of the strong interactions.

13 Coloring in the strong forces

While we began to understand how to construct renormalizable theories for the weak intractions, the *strong* interactions were still shrouded in mystery. They looked much less controllable. Investigators did learn how to cause various particles to collide with each other with increasing collision energies, using new and ever more powerful acceleration machines, resulting in a much better understanding of the curious internal structure of the hadrons. What was immediately obvious was that the list of resonances became longer and longer. A resonance could best be described as a particle that in all respects resembles one of the particles of Table 1, only its mass is bigger and often its spin is greater. The resonances came in series, with the heavier ones often having the highest spin. There are baryonic and mesonic resonances. Without changing the total of the quantum numbers S (strangeness) and I_3 (isospin), these resonances can decay into lighter particles within some 10^{-23} seconds.

It was difficult to understand why the hadrons behaved in this way. But, without attempting to answer such questions, Gabriele Veneziano discovered a simple mathematical formula that represented, in a particularly elegant way, the effects of all these resonances when particles collide. The remarkable thing about his formula was that in it the effects of the strong force were described very realistically (that is, relatively well in agreement with what was known from experiments), whereas no single existing theory could explain the formula. Not yet. Veneziano's formula would play a very important role, and I will return to it later.

If we further increase the collision energy, then so many resonances are produced (immediately decaying back into ordinary particles) that it becomes impossible to distinguish between them. But then another striking phenomenon is observed. Imagine that the strongly interacting particles (hadrons) are made from some jelly-like substance. If they are thrown together with great force, you would expect that, while destroying each other completely, the bits and pieces would continue more or less in the same direction, that is, most collision products would be expected to go in the *forward direction*. To be sure, this is indeed what often happens. However, investigators found also that, every now and then, fragments are thrown with great force sideways. *As if there were very hard grains in the jelly!* James Bjorken discovered that the sideways-going ejecta then obey rather simple equations: the collision processes at very high energy and those at a little less energy resemble each other more and more. This was to be called *Bjorken scaling*.

Feynman was intrigued by this phenomenon. It should be possible to explain Bjorken scaling using 'ordinary' field theory. Just assume that the resonances consist of more fundamental building blocks. They do not have to be quarks, he cautioned; Feynman called them 'partons'. If you assume that inside a resonance these partons behave more or less like free particles, then you could understand Bjorken scaling.

Quarks or partons, freely moving or impossible to shake loose? The conundrum was complete. I did not understand anything of this and decided to stay far away from the strong force. Ironically, I already had the answer to the Bjorken scaling problem in my note books! I had studied massless Yang–Mills theory well, and I had noticed something to which I first paid little attention. It concerns the 'scale transformation' discussed in Chapter 1. Remember, one can mimic the world of the large objects in the world of the small. Some details will look different, others remain the same. We find something like this in the world of elementary particles. If you look at a theory through a magnifying glass, you recognize the same particles as before. In particular, if you had no masses in your theory, the magnified particles are very much the same as the originals. Only if you apply renormalization does something happen. The naked particles then differ slightly from the dressed particles, and consequently the interaction strengths among the particles change a little when they are viewed through a magnifying glass (see Figure 13).

In the older theories physicists were used to, such as quantum electrodynamics and the Gell-Mann–Lévy model, the interactions in the magnifying glass *always* seem to be stronger than they originally were. The naked particles react more strongly than the dressed particles because they are covered by a blanket of 'vacuum particles'. These arise from 'vacuum polarization', and they tend to partly screen the charges. This turned out to be so in *all* field theories, *except* the new theories: the Yang–Mills gauge theories. There things work the other way, but since these theories were new and not so well studied, investigators had been unaware of this fact.

In a somewhat mysterious way, the naked particles are being surrounded by

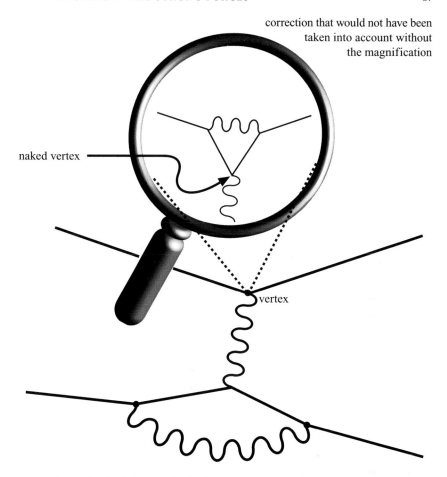

correction that would not have been
taken into account without
the magnification

naked vertex

vertex

Figure 13. The definition of a fundamental interaction depends on the magnification scale.

Yang–Mills photons from the vacuum in such a way that they reinforce the charge instead of weakening it. If you do this calculation for the first time it is a rather involved one, but I was aware of this phenomenon in 1972. Furthermore, I knew that if, besides Yang–Mills photons, the theory also contains fermions, then these tend to work in the other, the usual, direction; that is, they screen. If there are more than sixteen fermion types (this number is different for different gauge theories), then the vacuum blanket works in the same way as in the usual theories.

In 1972, a small conference took place in Marseille. On arriving at Marseille airport, I discovered that Kurt Symanzik and I had shared the same plane. We began to talk physics, and he explained his problem to me. He had been trying to understand Bjorken scaling in a quantum field theory, but had limited himself to what he considered to be the prototype of all field theories, a simple spin 0 model. Unfortunately, it had the wrong scaling behavior. 'If only I could turn this scaling behavior around', he said, 'then you would get a theory where particles at close distance behave almost like free particles, but when they separate to larger distances they would feel much stronger forces'. 'Well', I cried, 'that is exactly what you get in a Yang–Mills gauge theory!' Symanzik found this difficult to believe. In his opinion, I must have made a sign error somewhere in my calculations. There were theorems, 'no-go theorems', that stated that *all* theories scale like the spin 0 model!† 'And if that isn't true, if you are right, you should publish this quickly, because this would be very important!'

Much to my later regret, I did not follow this sensible advice. I was very busy with a calculation concerning quantum gravity with Veltman, and besides, I had used very special methods for my calculation in order to bring it into decent proportions, and it would have taken me some time to write all these methods down. I did not know how I should present these calculations; after all, I did not really understand the strong interactions. Now I am grateful to Symanzik. He terminated his lecture in Marseille with the words: 'It seems that no theory at all can explain Bjorken scaling, but perhaps we should have another look at these gauge theories.' And so he practically forced me to stand up at the discussion session to remark that I had done this calculation, and I wrote the result down on the blackboard.

The first written publications on the peculiar scaling behavior of gauge theories came from David Politzer, a young researcher at Harvard, and simultaneously from David Gross and Franck Wilczek in Princeton. They called this feature 'asymptotic freedom'.‡ It was Symanzik who later proclaimed to anyone who listened that this result had been obtained earlier in Europe, and that my public remark made at the Marseille conference qualifies in questions of priority.

This had been a missing piece in Nature's big jigsaw puzzle. Other pieces had been laid before. Together with Harold Fritzsch, Gell-Mann had already pointed out that Yang–Mills forces were ideally suited for describing the binding

---

† These 'no-go theorems' were simply wrong. The possibility of ghost particles, cancelling out in the final expressions, had been overlooked.

‡ Later, at a large international conference, J. Iliopulos would remark: 'And like always when someone talks about freedom, what they really mean is something altogether different!' His country at that moment was governed by a military junta.

forces between quarks. The resourceful Japanese scientist, Yoachiro Nambu, also had this idea. It enables one to understand exactly why either three quarks or one quark and one antiquark can stick together, and also why the spin $\frac{3}{2}$ baryons form a decuplet and those with spin $\frac{1}{2}$ an octet.† The Yang–Mills fields responsible for such a force have *eight* electric as well as magnetic fields (another example of the eightfold way).

The charges that the quarks have are more complicated than ordinary electric charges, where we have either positive or negative, which can neutralize each other. Every quark species (*up, down, strange,* etc.) can have a 'color': red, green or blue. Quarks with different colors attract each other, forming lumps of matter with some color mixture. The only lumps of matter that can roam around freely in Nature are such a mixture of quarks that they have become 'colorless' (white or some shade of gray), according to a rule that approximates to: red + green + blue = white.

The antiquarks have the *conjugated* colors: magenta ('anti-red'), violet ('anti-green') or yellow ('anti-blue'). The Yang–Mills photons acting on color are called *gluons*, for obvious reasons. They carry both a color and an anticolor. This would allow for nine combinations, but one mixture, a superposition of red/anti-red, green/anti-green and blue/anti-blue, can be called colorless and therefore does not participate. So there are eight gluon species left.

Quarks, antiquarks and gluons together comprise Feynman's partons. So, how about introducing a Higgs–Kibble mechanism? In the early days, many physicists thought of constructing a theory for the strong force much like the one for the weak force; the 'naked' gluons then have no rest mass. Such massless vector particles do not exist in the real world; there is only one ordinary photon. So they should get a mass via a Higgs–Kibble mechanism. However, wait! Particles *with* mass and other properties as expected for the gluons (such as their colors) do not exist either! They are forbidden since they are not gray.

The answer had to be something else this time. Now we know that for the strong force there is no Higgs–Kibble mechanism at all. There is a principle that forbids any particle or conglomeration of particles of which the total color is not completely neutralized. If from some colorless broth you try to isolate a particle

---

† There had been a difficulty in the old quark theory. This was that, in spite of the fact that quarks have spin $\frac{1}{2}$, they did not seem to behave like fermions. In the decuplet, for instance, the omega minus contains three identical quarks, all with their spins rotating in the same direction. That should have been forbidden by Pauli's exclusion principle! However, in the new theory, each of the quarks inside a particle such as the omega minus have a different color, and then one can show mathematically that they *have* to form decuplets and octets exactly as we see them in the baryons.

or whatever that has color, the leftovers would together have the conjugated color. The attractive force between color and conjugated color is so great that the amount of energy needed to completely separate them would be *infinite*. The force itself is not infinite, but does not decrease much with distance, unlike most ordinary forces. This is why such an isolation process will never be possible. In 1973 we understood very little of the deeper mathematical reasons behind this principle, but it was not difficult to think of a name for it: 'permanent quark confinement'. Our next task was to understand and explain this confinement phenomenon.

Now let me return to the marvellous formula proposed by Gabriele Veneziano, which I mentioned at the beginning of this chapter. The formula assumes that the resonances come in series: the heavier (hence more energetic) particles in such a series have more spin, so they rotate more wildly about their axes than the lighter ones. Exactly what this implies for the underlying physics seemed to be obscure, but then progress was made by the Danish physicist Holger B. Nielsen together with Ziro Koba, at the Niels Bohr Institute in Copenhagen, and others such as Yoshiro Nambu in Chicago: just suppose that these resonances also consist of something like 'quarks'. The large amount of rotational energy (or, more precisely, angular momentum) of the heavier resonances can then only be understood if the quarks in these resonances orbit in larger circles around each other, since each of them is not allowed to go faster than light. And yet the quarks never become separated from each other. The Veneziano formula is exactly what you get if you imagine that these hadrons consist of little pieces of 'string', with quarks at their end points. In the resonances the string rotates, or 'resonates'. The strings themselves seem to be made of some superstrong material. They are always very tightly strained, so that it takes a lot of energy to stretch them. If you do manage to stretch a string, the energy you are spending is immediately converted into more string material. And so you can stretch the string indefinitely. It becomes longer and longer and does not weaken at all! If quarks *are* bound together with such ideal brands of elastic strings, it is little wonder that they cannot be separated.

Nothing is ideal. If you stretch a string very far it does not weaken, but something else may happen. It can snap! But it does not snap without creating a new quark and a corresponding antiquark at the new end points. End points without quarks or antiquarks do not exist. And, if there is a quark at one end, there is always an antiquark at the other.

Baryons, consisting of three quarks, are obtained by allowing that three string end points can be tied together in a special way. I will not bother you with

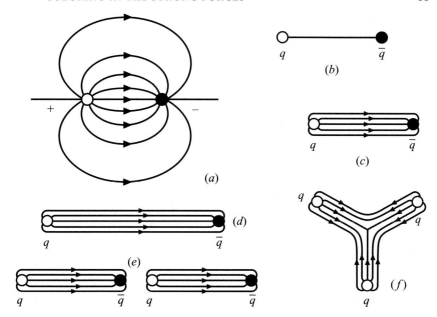

Figure 14. (a) Field lines between two charges in Maxwell's theory of electromagnetism. (b) String theory. (c) The gluon field between a quark and an antiquark. (d) On separating a quark and an antiquark more field is created, until (e) a new quark–antiquark pair is created. (f) The baryon. q = quark, $\bar{q}$ = antiquark.

the details of the rules. To visualize this one has to realize that a string has an *orientation*, one end point obeys rules that differ from those of the other end point. The *tension strength* of a string is a constant of Nature: about 14 *tons* of force.

At first sight, it seems difficult to reconcile this string theory with the theory that says that quarks are held together by gauge forces. But Veneziano's formula was only *approximately* right, it does not hold exactly. It is natural to suspect that the description in terms of strings with quarks at their ends is no more than an approximation. What we presently believe is that the *field lines* of the gauge field between quarks form a pattern as sketched in Figure 14(c). If you try to pull the quarks apart, the energy imparted to them is used to create more stretches of field lines between the quarks.

Perhaps you think our explanation is an acceptable and convincing one, and maybe you like the pictures in Figure 14. But there is an important problem. The calculations mentioned earlier, such as for the electron's magnetic moment, the renormalizability of the weak force, and the phenomenon of 'asymptotic freedom', could be carried out with great mathematical precision. You begin with a complete set of equations, which you have reason to believe describe something interesting (a 'model'), and you derive, using mathematical logic, the properties you want to know more about.

This was not at all possible with the confinement phenomenon. We could write down the equations: a Yang–Mills theory with color photons and fermions, that is, quarks, exactly following the rules of Chapter 12. We could demonstrate that the theory is renormalizable and asymptotically free. The latter implies that the forces become relatively weak when the quarks come very close to each other. Conversely, if one tries to separate the quarks further, the forces become very strong when compared with the ones inherent in the theory of electromagnetism. So far, everything was well established. But why do the force lines behave like strings? Why is the energy needed to separate the quarks infinite? For particles that stick together by the electric force (such as electrons in atoms, or ionized atoms in molecules) this would be the ionization energy. So why is the ionization energy for quarks infinite? The diagrams of Figure 14 are believable, but are they correct? Do they *follow on* from the equations we believe in? Unfortunately, precisely because the forces become strong, our calculations become unreliable.

This is the quark confinement problem. As I see things now, the resolution of this problem will be in two parts. First, we have to find a general scenario, an accurate but qualitative description of this peculiar property of the theory. We are dealing here with a *phase transition*, comparable to the transition from the solid phase into the liquid or the gaseous phase of ordinary matter, or into the superconducting phase or the Higgs–Kibble phase. The *confinement phase* is a state in which the photons do not have a mass, do not screen the charges like in the Higgs–Kibble phase, but where they bind all charges with infinite efficiency. In this sense, we would soon come to understand quark confinement very well, and I will return to this subject later.

Secondly, we have to deduce whether this 'confinement phase' is indeed realized, and, if it is, under what conditions. This question is a lot harder than the first one; it is to be compared with the question of which materials will become superconducting at low temperatures and which will not, or which materials are liquid and which are solid, and at which temperatures? In principle, all this should follow from the fundamental forces between the atoms or the elementary particles. The answer will depend on extremely technical calculations, for which large computers are often indispensible. This cannot easily be done with pen and paper anymore, because the forces among the particles have become so great that none of the simple approximation techniques can be relied upon. Even the

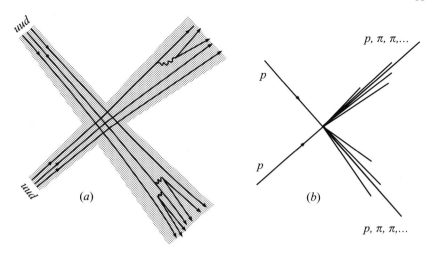

Figure 15. (a) Two protons collide with each other such that the quarks do not hit (a 'peripheral' collision). (b) What an experimenter then sees.

calculations we do now, with the very largest computers, give outcomes that are not at all yet very accurate, but all the results indicate very strongly that permanent confinement is exactly what happens in a color gauge theory.

In around 1974, the idea of a pure color gauge theory for the strong force, *without* any spin 0 particles, hence without any Higgs–Kibble mechanism, gained more and more support. The theory was called *quantum chromodynamics*, after the Greek word $\chi\rho\tilde{\omega}\mu\alpha$ = color.

This theory dawned upon the researchers very slowly. Not one person made this discovery, but it was a cooperative enterprise. This is a big contrast with relativity theory and quantum mechanics, but no less revolutionary! The theory changed the picture we had of the strong force completely! Now, instead of complete arbitrariness, we had only one fundamental constant of Nature that was not given *a priori* by the theory. Then, all that is required is to give the values of the masses of the quarks. Only if the quarks sit snugly together does this mass have some noticeable effect; otherwise these parameters are relatively insignificant. This implied that all properties of the hadrons should now be calculable. Only a few years before, nobody present at our Cargèse summer school could have dreamt of such a powerful theory.

That the theory gained popularity only slowly was understandable. Calculations were possible *in principle*, but it would take years before very powerful computers gave some results about the hadron masses that were at all encour-

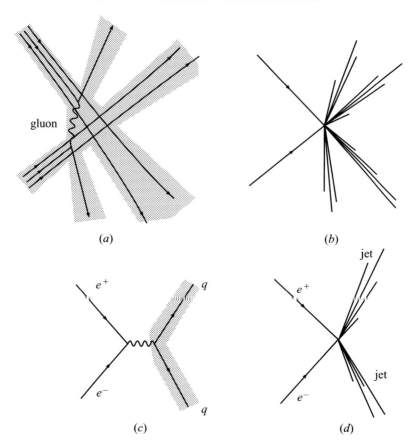

Figure 16. (a) Two quarks hit each other. (b) What the experimenter then sees: a typical 'jet event'. (c) $e^+ + e^- \rightarrow$ two quarks. (d) How this event looks in an experiment.

aging. Predictions were made, for instance, that the strong force would manifest itself merely as a weak force when very tiny distance scales were considered. But how could this be checked in an experiment?

To justify our present enthusiasm, let me run ahead of my story. If particles are fired against each other with energies much *greater* than the masses of ordinary subatomic objects (the hadrons), then these hadrons behave like spongy snowballs in which the quarks wander about as tiny hard pebbles. During most of the collisions, these quarks just pass each other by and the snow disintegrates. The quarks (and gluons) that continue straight on possess most of the energy. These are the usually less interesting collision events; see Figure 15.

Every now and then, however, a collision is seen in which two quarks hit each other. The quarks proceed sideways, trailing other quarks and gluons along so as to neutralize their color. In this way, one obtains two (sometimes even more) compact clouds of hadronic particles that follow the new trajectories of the original quarks. We call these clouds *jets*. What the experimenter then sees is sketched in Figure 16.

The theory of jets was pioneered by George Sterman and Steven Weinberg in 1977. In other experiments, the quarks and gluons also manifest themselves as 'jets'. For instance, if you fire an electron against a positron, these two can annihilate each other while emitting a photon, which in turn produces a quark–antiquark pair (Figure 16(c)). The resulting quarks again form clouds of hadrons: two jets (Figure 16(d)). Sometimes the primary collision produces a third or fourth primary particle (mostly gluons). These too manifest themselves as jets. In short, the original quarks and gluons produced in a collision event betray their presence by virtue of the fact that they are disguised as jets. By measuring these jets accurately, the theory may be checked directly. In particular, it is now possible to measure the interaction strength between quarks and gluons.

Every time a new and more powerful particle accelerator is switched on, a new and higher energy domain is disclosed, and so we are able to measure the quark–gluon interaction strength at increasing energy. The higher the energy, the closer the original quarks come together. And, in complete agreement with 'asymptotic freedom', we see that this interaction strength diminishes ever further the higher we go in energy.

14 The magnetic monopole

The problem of permanent quark confinement became more and more intriguing to me during 1973 and 1974. What we knew was only that the color forces on the naked quarks at small distances are relatively small compared with the forces between dressed quarks at larger distances from each other. This was the 'negative screening effect' I mentioned in the previous chapter. It seems plausible that if this is extrapolated to larger distances, the color forces among quarks will continue to increase. The forces could easily become so strong that they keep the quarks permanently together. But is this really what happens? The essential difficulty was that because the forces are so great, calculations are nearly impossible. The only thing we are good at is performing calculations on particles that move in approximately straight lines. This is certainly not what quarks do when the forces are so strong. So the question became: is there a fundamental reason why quarks will remain invisible forever? Why do the field lines in a color theory form sausages, as in Figure 14, without spreading apart as in 'ordinary' electric and magnetic fields? If this could be explained, perhaps a calculational scheme could be devised to explain the properties of hadrons. I tried all sorts of things to find an answer.

But Holger Nielsen and Poul Olesen from Copenhagen, and independently Bruno Zumino at CERN, discovered something I had not thought of at all. Consider a piece of superconducting material, they said. Now try to put a *magnetic monopole* in there. Yes, we know that these may not exist, but we can always imagine them, and then calculate what the superconductor would do. A magnetic monopole is a particle that, unlike an ordinary magnet, possesses a magnetic north pole but not the corresponding south pole (or vice versa). Monopoles have never been discovered, but more about that later.

I explained earlier that a superconductor does not tolerate any electric or magnetic field: it screens the fields off completely. But it *cannot* do that with the field of the monopole once it has one completely inside. To admit a monopole the superconductor must create a small region where the material is no longer superconducting, and this becomes the region where the magnetic field lines of the monopole may go. The superconductor seeks to do this with the least expense of energy. Calculations show that the field lines between a magnetic north pole and a magnetic south pole take on precisely the sausage shape shown in Figure 14. Within the sausage, the material is no longer superconducting. However, because creating this region costs energy, it is kept as small as possible, which is why the field lines do not spread out (see also Figure 17(a)).

And so Nielsen, Olesen and Zumino obtained a model for quarks. Quarks are kinds of magnetic monopoles, and our space is a kind of superconductor. We have seen all of this before: a Higgs theory!

Unfortunately, it would be hard to defend the notion that quarks are really magnetic monopoles. According to quantum chromodynamics, in which, by 1973, I firmly believed, quarks have nothing at all that resembles magnetic charge. And just suppose that we simply forget quantum chromodynamics for a moment, and take up the theory of magnetic monopoles; at that time, no single useful theory existed that described the motions of magnetic monopoles accurately, because magnetic monopole charges always had to be *strong*. If all that charge is put into a point-like particle such as a quark, the forces among the particles would *always* be very strong, even at very close distances: no asymptotic freedom. In such a theory, nothing can be calculated anymore.

But the idea fascinated me. Let's look, I thought, at what happens if you do this for a Yang–Mills theory. And then I hit upon a crazy difficulty. I could argue that the 'flux tubes', as we began to call the sausage shaped field configurations, would become unstable. But how could that be? Magnetic field lines of a magnetic monopole can only end where we have another monopole. If the flux tube disappears, then where do the field lines go? And then I realized something very important from this Gedankenexperiment. The only possible reason for the flux tubes to decay in the model I was considering would be that other magnetic north poles and south poles were formed spontaneously. That such a thing could happen in a gauge theory was not yet known! This was something new and important.

Forget the quark confinement problem. The model I was looking at had nothing to do with the strong force. Unintentionally, I had returned to the weak force theory. In 1974, there were still several varieties of these littering the literature. I was now able to derive that some of these models contained not only the particles they were intended to describe, but in addition genuine magnetic monopoles. In particular, a model proposed some time earlier by Glashow and a younger researcher with him, Howard Georgi, would contain monopoles, constructed simply using the particles and fields in the model as building material. This was something entirely different from the older theories for 'magnetic monopoles', in which the existence of point-like monopoles in addition to the other known (point) particles simply had to be postulated. But my monopoles were not point-like at all. They were soft, spatially extended objects. I could calculate all their properties, including their mass, which had never been possible before. That all this had not been realized before was partly because the energy, or mass, of these new objects is much larger than that of 'ordinary' particles. This was an interesting finding, which I announced at an elementary particle conference in London. It turned out that the same discovery had been made in Moscow by Alexander Polyakov, in a discussion with Lev Okun.

The fundamental possibility of loose magnetic pole particles had previously been suggested by Paul Dirac in 1931. He had proved, on very general grounds, that the *product* of the fundamental electric charge of an electron and the fundamental magnetic charge of a possible magnetic monopole would always have to be an integer multiple of Planck's constant. This is somewhat inconvenient if you want to construct a model for magnetic monopoles, because it follows that the unit of magnetic charge for monopoles is 68.5 times as strong as the unit of electric charge (this number is exactly half of the famous number 137.036...; see the box entitled 'The electromagnetic force' on page 19). If quarks were monopoles, they could never be asymptotically free!

Of course, there had been numerous investigators who had nevertheless tried to treat magnetic monopoles following the usual perturbation approximations, with very discouraging results. Now, my monopole was not a point particle but a bunch of fields. A perturbative approximation that would only be valid when the magnetic charge was tiny was not needed here, and that is why my monopole† is so much nicer. It was important that I could calculate its mass precisely because this had never before been possible. If the Georgi–Glashow model was right, the magnetic monopole would be about 8000 times as heavy as the hydrogen atom.

Benjamin Lee was very impressed. While I was discussing this work with him in London, someone overheard and asked me in a somewhat surprised fashion: 'Hey, are you proposing the umpteenth model for magnetic monopoles?' Lee said: 'No, no, he does not have a model, he *found* one!' I adored Benjamin Lee. It was a great loss for elementary particle physics when, only a few years later, a tragic traffic accident took him away from this world.

---

† I say 'my monopole', but clearly it also belongs to Polyakov and Okun.

The Georgi–Glashow model contained magnetic monopoles, but unfortunately this was not the case for the Weinberg–Salam–Ward model, in which there is a technical obstruction that does not admit my monopole. This has to do with the neutral current, which was absent in the Georgi–Glashow model. And, as you already know, neutral currents were being detected (they were a big issue at this London conference!). So the Weinberg–Salam–Ward model was gaining territory, leaving the Georgi–Glashow model and others behind. If magnetic monopoles existed at all, they would presumably have to be much heavier than the 8000 proton masses that would have followed from the Georgi–Glashow model.

This theoretical development prompted experimentalists to search with re-newed energy for magnetic monopoles, in sea water, in lunar rocks, and in oysters.† Very sensitive detectors were built with superconducting coils. One single event seemed to be registered, but when the apparatus was refined and the detection efficiency improved, the signals stayed away. This is a well known phenomenon in experiments, indicating that the first signal (which had been called unreliable also by the physicists who first saw it) must have been based on an incorrect interpretation.

We still do not know whether magnetic monopoles exist, but the discovery of this theoretical possibility would remain important for two reasons. As we will see later, it is reasonable to assume that the Weinberg–Salam–Ward model only describes correctly those particles that can be detected experimentally by our machines in our time. But there is a limit to the energies one can give to a single particle in a laboratory. At energies very much beyond this limit, new phenomena are to be expected that can only be understood if we extend the Weinberg–Salam–Ward model further. We do not yet know what the improved theory will look like, but there is a reasonable likelihood that the ''t Hooft–Polyakov monopole' will then emerge as a solution. The model I will describe in Chapter 22 contains real but superheavy magnetic monopoles.

'Whatever is allowed is compulsory', some theoreticians claim. This means that every theoretically imaginable construction must exist somewhere in Nature. It is not always true, I am afraid, but it does hold for magnetic monopoles. *If* they are theoretically possible, one can also calculate how many of them will have been produced in the very early stages of the universe. The results of such calculations cause headaches to those theoreticians who wish to understand the early phases of the cosmos. All those monopoles would have made our universe look quite different from what we see at present. In fact, these cosmological calculations

† Oysters are known to accumulate strange minerals from sea water.

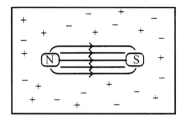
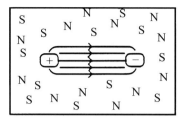

Figure 17. (*a*) In a superconductor there are positive or negative charges that collectively populate the lowest energy level ('Bose condensation'). The magnetic field between a north monopole (N) and a south monopole (S) is forced to take the shape of a sausage ('vortex'). (*b*) In quantum chromodynamics we have colored magnetic monopoles that condense to populate empty space. Because of this it is now the color *electric* field between two quarks that is forced into a vortex.

are already being used to derive restrictions on those particle models that would admit magnetic monopoles.

The second important point concerning monopoles was that we came to realize that quantum chromodynamics would also admit color magnetic monopoles, and even large quantities of them at that. If one assumes that these color magnetic monopoles undergo a kind of Bose condensation, the beginning of a more realistic explanation of the quark confinement phenomenon in quantum chromodynamics is touched upon: we use the original idea of Nielsen and Olesen (Figure 17(*a*)), but we exchange all electric charges and fields with the magnetic ones. The quarks are color *electric*; the surrounding vacuum is a color *magnetic* superconductor! See Figure 17(*b*).

This is difficult but fascinating material. With the color magnetic monopole, another piece of our Grand Jigsaw Puzzle seemed to fall into place. The Nielsen–Olesen solution and the magnetic monopole would turn out to be merely the beginning of a series of remarkable mathematical features in Yang–Mills theories, which are still keeping mathematicians and physicists busy to this day.

15 Gypsy

In November 1974, the elementary particle world was rocked by a surprising discovery. I was visiting Paris when the birth was announced of a new particle. 'A new particle?', you might wonder. 'And there were so many already. What's the big deal?' Well, it turned out to be a particle that did not fit into any of the existing series. Two groups of experimenters had made the discovery independently.

Samuel Ting was leading an experiment in Brookhaven, near New York, in which very high energy protons were made to collide with a target made of heavier material. Over several months, he and his collaborators had been observing a curious 'signal' in their apparatus. Ting found it difficult to believe that this signal was to be identified as a new particle, because, if this were true, it had to be something really spectacular. He went off to check and double check (including the possibility that he was the victim of some practical joke), and he ordered complete secrecy from all his collaborators.

The new particle, which he called J, would decay extraordinarily efficiently into an electron and a positron. It was these two particles that Ting detected in pairs, and when he measured their relative energies, it was found that they apparently originated from a new chunk of material with a mass of 3100 MeV (more than three times the proton mass). The laws of particle physics tell us that it then should also be possible to produce this object by colliding electrons and positrons head on with energies of 1550 MeV each.

Near Stanford, California, there is a laboratory named SLAC ('Stanford Linear Accelerator Center'). In a new accelerator, SPEAR, electrons and positrons were smashed against each other. When Ting heard that the SPEAR machine was being tuned to around 1500 MeV per particle, he realized that he should make his result public, but by then it was too late: suddenly the counters in the detectors at SLAC went wild!

If the electrons and the positrons are each given 1540 MeV, or 1560 MeV, they act as if they do not see each other. The beams of electrons and positrons go right through each other, causing practically no disturbance. The physicists had become used to expecting collision events occurring about once every minute. But if they tuned their machine to 1550 MeV/particle, suddenly a lot of direct hits were produced , on average one every second. New particles were being produced that immediately decayed again, leaving signals in the detection apparatus. For the leader of this experiment, Burton Richter, it was very soon clear that such a drastic increase of the collision frequency could not be blamed on a bug in the machine. He announced the discovery of a new particle, which he called Ψ (psi, after the first two letters of SPEAR). The wild clicks of the detectors were amplified and passed on to the laboratory's intercom, so that all the SLAC employees could enjoy the new discovery. Only after this did they hear about Ting's results. Not much was needed to convince Ting and Richter that they had both made the same discovery.

What was so curious about J/Ψ (or, more colloquially, 'gypsy') was that it did not seem to belong anywhere in the particle tables. It could decay in very many different ways: $e^+ + e^-$, $\mu^+ + \mu^-$, 5π, 7π, 9π, $2\pi + 2K$, and, though less often, into other numbers of pions, kaons and baryon–antibaryon pairs. Sometimes a photon also emerged. In the Paris laboratory of the École Normale Supérieure where I was, one member of the SLAC team came to a packed lecture hall to explain what they had learned about the new particle. He ended with a list of possible theoretical explanations of the observations, and how one could identify the new particle.

Was it a meson? Its average lifetime, (10^{-20} seconds) was much too long for that. Is there a conservation law holding back these decays? But then whatever force *does* give rise to the decay must be much stronger than the weak force. Was the new particle a new kind of photon? There were some indications in favor of this view. It seemed likely that the spin equalled 1, just as for the ordinary photon. Was this then perhaps one of the Yang–Mills photons of the weak force? But for that its mass was far too low. Or was it the carrier of the strong force? If that were true, absolutely nothing could be right about what I thought I knew about quantum chromodynamics.

One suggestion was that J/Ψ was a bound state of objects we had long sought for, a *charm* quark, c, and its own antiparticle, $\bar{c}$. A very illuminating thought crossed my mind. 'Hey', I said to my colleague sitting next to me, 'That's a neat possibility. What's wrong with that?' But his answer was what many of the other theoreticians present said: 'Nah, then it would be an ordinary meson, and for that its lifetime is much too long!' c and $\bar{c}$ could annihilate each other using the strong force; what prevents them from doing that for so long?

We had barely recovered from the shock, when we were told that there were more of these kinds of particles. Ψ' has a mass of nearly 3700 MeV, and Ψ'' weighs 3770 MeV, but this latter was much less stable, behaving more like a good resonance should. Had we found more colored bosons here, or was this the beginning of a new series of resonances?

Now, a bit more about my brainwave. I cannot really make any priority claim here, because the confirmation that J/Ψ really is a $c\bar{c}$ meson came from further careful experimentation as well as theoretical analysis done by various theoreticians closer involved with the experiments than I was. They were exhilarated because this was the clearest signal yet that charm really existed. The reason why J/Ψ or $c\bar{c}$ is so curiously stable is due to asymptotic freedom. Why?

The c quark is much heavier than the quarks we discussed earlier, u, d and s. In a $c\bar{c}$ bound state, both quarks move much closer to each other than 'normal' quarks would do – the typical distance over which a strongly interacting object would vibrate is inversely proportional to its mass. So the $c\bar{c}$ 'meson' is much smaller than ordinary mesons. You may remember that the strong force becomes less strong when it acts over a smaller distance, and this is due to the negative charge screening effect (Chapter 13). It implies that all strong interactions inside the $c\bar{c}$ system proceed unusually slowly, and that is why this meson lives much longer than its cousins $s\bar{s}$ (the phi resonance) and $u\bar{u}$ and $d\bar{d}$ (rho and omega).

The bound state of an electron and a positron looks a bit like a curiously light 'element', and was therefore called 'positronium'. Because of its close analogy, the $c\bar{c}$ bound state was soon called 'charmonium'. An extra consequence of the fact that for charmonium the strong force is relatively weak is that many calculations could now be carried out with much greater accuracy than we were used to in other strong interaction processes. The charmonium system was therefore the first system allowing us to carry out reliable quantum chromodynamics calculations! In fact, the calculation of its lifetime had to resemble very closely the one for the positronium lifetime. Positronium can decay into photons via the electromagnetic force. To calculate the charmonium lifetime, all you had to do was to take this calculation for positronium and replace the electric charge in there by the strong interaction coupling constant.

The calculation itself is complicated. You would not be able to do it on the back of an envelope. As soon as I returned home, I looked up positronium in the textbooks. There I saw something I had not even thought of. If you take the spin 1 bound state, it cannot decay into two photons, there have to be at least three. This in turn implied that the decay probability would, in any case, have to go as the sixth power of the strong coupling constant. The strong interaction

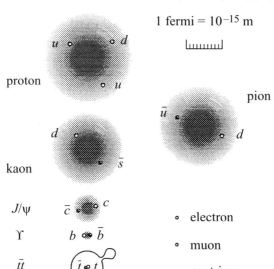

Figure 18. Particle sizes.

strength for the $c\bar{c}$ system had only to be a little weaker than that of the $s\bar{s}$ or $u\bar{u}$ systems to explain the long lifetime of J/Ψ!

The J/Ψ particle decays when the two quarks annihilate each other by first producing three *gluons.* These gluons then in turn create pions, kaons or whatever, but because the first interaction goes very slowly it determines the total lifetime. On substituting the numbers, it was found that they seemed to fit the observations reasonably well. It did not take long for the particle community to realize and accept this explanation.

The charmed quark also had to be able to present itself without its own antiparticle, accompanied only by ordinary quarks. In particular there had to exist mesons of the type $c\bar{u}$, $\bar{s}c$, etc., and also baryons made of the combinations udc, usc and others. As for the existence of such hadrons, several signals had already been detected, in the form of registrations of unusual collision processes. But details were scarce.

Now that the c quark's properties could be determined with much more accuracy, the experimenters also knew precisely where they could find more of them. On computing the masses of these particles, it seemed reasonable to deduce that the heavier Ψ'' could decay directly into charmed mesons. Since this does not require that the charmed quark and antiquark first annihilate each other, this decay happens much faster than that of J/Ψ itself. This is why Ψ'' is much less stable. But this is also why the Ψ'' decay products are the ideal place to look for the charmed mesons, and sure enough they were soon discovered there.

Was the J/Ψ, just like the muon, not ordered by anybody? In hindsight, it could be seen that there had been attempts to try to detect $c\bar{c}$ mesons, but the consequences of asymptotic freedom in the colored force had been underestimated. These mesons had been expected to look much more like the resonances ρ (rho) and φ (phi), which consist of ordinary quarks and antiquarks, also in the spin 1 state. Now we were banging our heads against the wall: we could have predicted everything about the J/Ψ!

J/Ψ was a piece that tied together various loose parts of our Grand Jigsaw Puzzle. On the one hand, we had the missing charm quark, needed to cure the weak force theory, via the GIM mechanism; on the other hand, quantum chromodynamics, with its asymptotic freedom, turned out to behave so decently in accordance to the theoretical rules that we had ourselves been taken by surprise! Apparently, quantum chromodynamics was not just a simplified model: the details, which only few of us had taken seriously until then, closely matched the observations all the way!

Veltman was, typically, skeptical. If you have a theory, he said, you should make a prediction, and not say in hindsight that everything fits. I was prepared to make a prediction. Since the forces in J/Ψ are weaker, the quarks in there should move considerably slower than the speed of light. Therefore the effects of the *spin* of these quarks should be relatively unimportant. In J/Ψ the two quarks rotate in the same direction, which is why in this combination the total spin is 1. But they could also rotate in opposite directions, so that a spin 0 state is obtained. This would be again a new particle, and experimenters had already started to look for it. It is much more difficult to detect than J/Ψ (it cannot be made directly by hitting electrons against positrons). My prediction was that its mass would only be slightly lower than that of J/Ψ. For the ordinary quarks the mass differences are relatively high. I believed that the new, yet to be discovered, particle should have a mass between 3000 and 3100 MeV. Most other physicists expected a lower mass.

This became a bet, which I lost. The new spin 0 particle, which was to become known as η_c, the 'charmed version of the eta (η) meson', would turn out to be only 2980 MeV, just below the limit of my bet. As I mentioned above, η_c cannot be made by colliding electrons against positrons. Also, it cannot decay into e^+ and e^-. The reason is that such a decay can only happen if first an ordinary photon is made, which can then turn into the electron–positron pair. But the photon has spin 1, not 0. The way experimentalists had to look for η_c was to wait for a J/Ψ to decay into η_c and a photon. This decay is rare, and if it does happen the analysis is not so easy. But the experimentalists are true artists when confronted with problems of this sort.

The reason why η_c is lighter than I had thought is that between these quarks, the strong force is not yet *that* weak. Therefore, deviations from the straight positronium calculations are already considerable. My bet had been a little too impulsive.

My elderly colleagues in physics often look back wistfully to 'those glorious days of physics'. What they are referring to is the era of the big discoveries in the first half of the twentieth century: quantum mechanics, general relativity, quantum electrodynamics and the discoveries of the first elementary particles. But for me the 'glorious days' were between 1970 and 1976, when so many pieces of the Jigsaw Puzzle for the weak, electromagnetic and strong forces all fell into place. The J/Ψ discovery in 1974 was the climax. Before that discovery was made, there was still some doubt as to whether we had the right weak interaction theory, and we were considering our strong interaction theory as nothing but an idealization of something that could well be much more complicated and unfathomable. Now we were suddenly sure that both these theories were right, even in their details. As experiments continued, our astonishment about this grew. The details were more accurate than many of us had ever hoped. One thing was clearer than ever: we live in a world that obeys meticulously the laws of mathematics. The mathematics is difficult, but it can be understood entirely.

Perhaps this is one of those wonders that invites a moment of thought. How can it be that our human brains have been able to understand this totally alien world of the most exotic subatomic particles in so much detail? Did not those brains of ours evolve according to natural selection: he who can manufacture and manipulate the best bow and arrow will be given the opportunity to pass this ability to his progeny? And what do elementary particles have in common with a bow and arrow? The only answer I can think of is that they indeed have *a lot* in common: logic is logic, and it applies to bows and arrows exactly in the same way as to elementary particles.

Measuring the lifetimes of elementary particles containing charm was a new challenge for the experimentalists. Many of these particles have lifetimes as short as 10^{-12} seconds, and because of that they can travel only a fraction of a millimeter during their entire lifetime. In this realm, working with wire or bubble chambers becomes difficult. But experimentalists are inventive. By measuring accurately, and above all patiently (sometimes a particle lives a bit longer), they still manage to measure the average length of a track. Sometimes a particle is sent through a photographic emulsion that can be further analyzed through a microscope.

16 The brilliance of the Standard Model

For the particles discovered so far, we now had a pretty detailed map of the weak, the electromagnetic and the strong interactions. Only one thing was not quite right in the picture presented so-far: on average, three out of every thousand K_{Long} mesons decay into just two pions, and thus they violate conservation of PC symmetry, as I described in Chapter 7. Which force is responsible for this phenomenon?

What makes this problem particularly difficult is that K_{Long} is the *only* particle in which this peculiar force has manifested itself up till now.† K_{Long} is composed of two quarks, *strange* and *down*. Apparently, quarks exert forces on each other that are not quite PC symmetric. Where could such a force come from? Until now, our mathematical scheme has not included such a force; all our equations have automatically been PC symmetric. Several ways of repairing our model, by adding a PC violating force, such that the K_{Long} decay could be explained, have been invented. In any case, more 'auxiliary particles' would be needed to transmit the new force.

It turns out *not* to be possible to repair our model by invoking yet another Yang–Mills field. The spin 1 particles always preserve PC symmetry (could this be why the violation of PC symmetry is so tenuous?)

One can imagine the effects that could come from another spin 0 particle, preferably also subject to some kind of Bose condensation. The result would be what we call 'spontaneous PC violation'. However, the resulting models then obtained were not very popular. We wish to avoid spin 0 particles as much as possible because they bring along so many arbitrary interaction parameters. Models with such particles look very artificial. By itself, such an argument is of

---

† Attempts to see PC violation in neutrons have not yet been successful. More promising are plans to look for PC violation in D and $\overline{D}$ or B and $\overline{B}$ mesons.

course not sufficient to exclude this possibility, but it so happens that there is a more interesting possibility.

You may remember that Glashow, Iliopoulos and Maiani had introduced the *charm* quark in order to understand the symmetry structure of the weak force. Well, it was proposed, let's do this again. This time, we need to introduce two more quarks. The first four had formed pairs (u and d, and c and s), with electric charges $+\frac{2}{3}$ and $-\frac{1}{3}$. The new pair had to look just like this, but the new quarks could easily be considerably heavier than the already known ones. Being analogous to the 'up' and 'down' quarks, they were named 'top' (t) and 'bottom' (b), respectively. But sometimes the same letters are used to give them the more poetic names 'truth' and 'beauty'.

The need for one spin 0 particle to give the weak force the symmetries it has via the Higgs–Kibble mechanism had been inevitable. This one Higgs particle now couples to the quarks and leptons to give them their masses. But the same Higgs particle can also produce transitions between the various quark types. If there were no weak force at all, the quarks could have stayed in all sorts of stable states. Now it is a *conspiracy* between the weak force and the Higgs force that enables the many types of decays of the strange and charmed hadrons.

The more fermions we introduce, the more kinds of interactions the Higgs field may have with these quarks. The Japanese physicists M. Kobayashi and K. Maskawa wrote down the most general mathematical expression for the forces one then obtains. It turned out that one of the terms in their equations is not PC symmetric. But such a term only arises if there are at least six quark species. This is why a search began for particles containing yet another quark species. And, sure enough, they were found. In Table 6, some particles are listed that have the new 'beauty' quark. Their discovery followed the same pattern as those particles with charm. In 1977, the Υ (upsilon) was found, consisting of a b and a $\bar{b}$. Its mass is as much as 9460 MeV. Then hadrons were found containing only one b quark, weighing more than 5000 MeV. Apparently this beauty quark costs about 5000 MeV of energy.

The sixth quark, called 'top', or 'truth', was the most difficult to produce and to detect. One of the main reasons was that its mass was very difficult to predict theoretically. Even though more and more powerful machines were being built, experimenters failed to detect anything for a long time. Hence, the lowest possible value for the mass still consistent with these results was a number that steadily rose. But physicists were confident that the 'top' quark would be found.

We arrived at a model featuring six quark species ('up', 'down', 'strange', 'charm', 'bottom' and 'top') and four lepton species (the electron, the muon and

Table 6. *Some of the new particles*[a]

Particle	Quarks	Mass (MeV)	Spin	Lifetime (seconds)	Some decay modes
J/ψ	$c\bar{c}$	3096.9	1	7.5×10^{-21}	e^+e^-, $\mu^+\mu^-$, 5π, 7π, $3\pi+2K$, $p\bar{p}\pi^+\pi^-$, etc.
Ψ'	$c\bar{c}$	3686.0	1	2.4×10^{-21}	$J/\psi+2\pi$, e^++e^-, $\mu^++\mu^-$, 5π, etc.
ψ''	$c\bar{c}$	3770	1	2.8×10^{-23}	D^++D^-, $D^0+\overline{D}^0$, e^++e^-
η_c	$c\bar{c}$	2979	0	6.4×10^{-23}	$\eta+2\pi$, 4π, $2K+2\pi$, $p+\bar{p}$
χ	$c\bar{c}$	3415	0	5×10^{-23}	4π, $2\pi+2K$, 6π, 2π, $J/\psi+\gamma$, $2K$
χ'	$c\bar{c}$	3510	1	7.5×10^{-22}	$J/\psi+\gamma$, 4π, 6π, etc.
χ''	$c\bar{c}$	3556	2	3×10^{-22}	$J/\psi+\gamma$, 4π, 6π, etc.
Υ	$b\bar{b}$	9460	1	1.2×10^{-20}	$\mu^++\mu^-$, e^++e^-, $\tau^++\tau^-$, hadrons
Υ'	$b\bar{b}$	10023	1	1.5×10^{-20}	$\Upsilon+2\pi$, $\mu\mu$, ee, hadrons
Υ''	$b\bar{b}$	10355	1	2.5×10^{-20}	$\Upsilon+2\pi$, $\Upsilon'+2\pi$, $\mu\mu$, ee, hadrons
D^+	$c\bar{d}$	1869	0	1.06×10^{-12}	$K^-+2\pi^+$, $K^-+4\pi$, $\overline{K}^0+\pi^+$, $\overline{K}^0+2\pi$
D^-	$d\bar{c}$	1869	0	1.06×10^{-12}	$K^++2\pi^-$, $K^++4\pi$, $K^0+\pi^-$, $K^0+2\pi$
D^0	$c\bar{u}$	1865	0	4.2×10^{-13}	$K^-+\pi^+$, $\overline{K}^0+\pi^0$, $K+2\pi$, $K+3\pi$, 2π, $2K$, etc.
$\overline{D}^0$	$\bar{c}u$	1865	0	4.2×10^{-13}	$K^++\pi^-$, $K^0+\pi^0$, $K+2\pi$, $K+3\pi$, 2π, $2K$, etc.
F^+	$c\bar{s}$	1969	0	4.7×10^{-13}	$2K$, $2K+\pi$, 4π, 6π, etc.
F^-	$s\bar{c}$	1969	0	4.7×10^{-13}	$2K$, $2K+\pi$, 4π, 6π, etc.
B^+	$u\bar{b}$	5279	0	1.5×10^{-12}	$\overline{D}^0+\pi^+$, $\overline{D}^0+e^++\nu_e$, etc.
B^-	$b\bar{u}$	5279	0	1.5×10^{-12}	$D^0+\pi^-$, $D^0+e^-+\bar{\nu}_e$, etc.
B^0	$d\bar{b}$	5279	0	1.5×10^{-12}	$\overline{D}^0+\pi^++\pi^-$, etc.
$\overline{B}^0$	$b\bar{d}$	5279	0	1.5×10^{-12}	$D^0+\pi^++\pi^-$, etc.
Λ_c^+	cdu	2285	$\frac{1}{2}$	2×10^{-13}	$\Lambda+\pi^+$, $p+K^-+\pi^+$, $p+\overline{K}^0$, etc.

[a] Numbers according to 1995 data.

the corresponding neutrinos). This model could explain everything we knew about particle physics in the early 1970s. Unfortunately, we also knew it could not be completely correct. It would still contain a tiny dirty spot – something that at first sight looks more like a purely technical, mathematical problem, and it would be amazing if Nature itself would care at all about our little technical problems. The fact is, however, that the renormalization procedure does not quite work as well as we would like if there are six quark species but only four leptons. We call this an anomaly. What exactly the *physical* reason is for this anomaly I will explain later.

Fortunately, it was rather simple to repair our theory. All we needed was a new pair of leptons, and the experimenter Martin Perl, at the Stanford Linear Accelerator Center, had already begun to look for them. Later, he explained why he thought there had to be more leptons; his reasoning had nothing to do with the theory just mentioned. The strongly interacting particles come in infinite series, protons, neutrons and then all their excited states (the resonances). Why should the *weakly* interacting particles stop at two pairs (the electron and the muon with their neutrinos)? Perl thought that there could be as many leptons as there are hadrons, an infinite series, but that they are just a bit harder to find.

The difficulties with detecting a new lepton are its short lifetime, combined with the fact that it does not decay without also emitting neutrinos, which are impossible to detect. Therefore the signal of a decaying new lepton is difficult to recognize. When Perl did find a signal, in 1975, he had a further problem: convincing his colleagues that it was a reliable signal, and that he had therefore identified a new lepton. I also was skeptical in the beginning, but now the measurements are so accurate that there is no further reason for doubt. The new lepton is called 'tau' (τ) and weighs 1784 MeV. It decays into $e^- + \bar{v}_e + v_\tau$, $\mu^- + \bar{v}_\mu + v_\tau$, $\pi^- + v_\tau$ or into other hadrons, always with a v_τ among the decay products. As you might imagine, Perl continued his search. 'I am right on track', he thought. 'Surely there will be many more out there'. But, to his disappointment, the series seemed to stop there, at least for the time being.

There is little doubt now that the weak force is correctly described by the model originally written down by Weinberg in 1964, with a few extensions. By 1980, the measurements of the weak currents had become so accurate that much more detailed predictions could be made concerning the carriers of the weak force, the charged W^+ and W^- bosons, and the neutral carrier, which Weinberg had called Z^0 (it would keep that name). W^+ and W^- were predicted to have masses of a little more than 80 000 MeV, and Z^0's mass is slightly greater than 90 000 MeV.

During the 1970 Cargèse summer school, I heard that a Dutch engineer had found a smart way to produce in an accelerator a very tight beam of antiprotons, which are made by bombarding atoms in a target with protons. The antiprotons are produced together with a lot of other particles, all of which shoot off in all directions. By letting the antiprotons go through an electric potential difference, they will all be pushed in the same direction, but the beam will remain very diffuse. What one would like to have is a beam of antiprotons colliding against a beam of protons, but this would only be possible if both beams were extremely sharply focused. For the protons this was not so much of a problem, but how do you focus the antiprotons if they do not all start off with the same speed in the same direction? It seemed to be mathematically impossible.

Yet it was not impossible at all. If particles move at random with large velocities relative to each other, we say such a system is 'hot'. If they move with small relative velocities, we say the system is 'cool'. So the question was, how do we 'cool' a beam of antiprotons? It was to this problem that Simon van der Meer at CERN found a neat solution. Imagine a class of undisciplined school children, who all do some different naughty things. There is only one school teacher who can only shout to the entire class. Every time he shouts some command, only a few obey. Some of the others react by becoming even worse. By choosing his words carefully, the teacher will gradually bring the class under control.

Now replace the school children by antiprotons, moving in more or less circular orbits. The eyes and ears of the school teacher become sensitive electrodes that register the average conditions of the beam. The commands are given by rapidly varying electric or magnetic fields. The result is that the antiprotons all come closer and closer to one common ideal orbit. This process is what Van der Meer called 'stochastic cooling'. After elaborate experimentation at CERN, it turned out that the method worked well. A super proton collider named SpS was rebuilt into $S\bar{p}pS$ (Super Antiproton–Proton Synchrotron). In near-circular tubes, protons and antiprotons were accelerated and cooled, after which they could be focused onto each other and the collisions could be studied. The collisions occurred with so much energy that W^+ and W^- particles could be produced. Indeed, after the required energies were reached, the fingerprints were soon observed of both the W and the Z particles.

Next, an even larger machine was built, called LEP ('Large Electron–Positron Collider'). This machine required a circular tunnel with a circumference of nearly 27 kilometers (more than 16 miles). Accelerating electrons and positrons to high energies is much more demanding than accelerating protons and antiprotons because of their much smaller masses (much more radiation is being emitted when these light particles are being accelerated). But the big advantage of studying electron–positron collisions is that these two particles together can produce exactly one Z^0 boson, without wasting their energies producing anything else. If the electrons and the positrons are given each a little more than 45 000 MeV of energy, then *large quantities* of Z^0 particles are made, and the sensitive energy dependence of this process enables an experimenter to carry out very detailed measurements. LEP is a huge success story. The machine works extraordinarily well, and shortly after its completion, in 1989, it began to produce heaps of Z^0's. The only disappointment for the theoreticians is that no new disagreements with the established theory were found.

Once the Z^0 particles were being produced in such great quantities, experimenters were able to perform precision measurements on them. Indirect effects that other particles have on the properties of Z^0's could be detected, and many mathematical details of the emerging theory could be verified. Even the effects of the still missing top quark could be accounted for, and this gave theorists a new way to estimate its mass. Its value was first estimated to be somewhere around 135 000 MeV, and was later corrected to the higher value of 175 000 MeV, give or take 20 000 MeV or so. A powerful machine near Chicago called the Fermi National Accelerator Laboratory, or 'Fermilab' for short, was soon going to produce the enormous energies required for the production of pieces of matter in this mass range. In 1994, two American teams could confirm the earlier results: they had found a top quark, and its mass, though still somewhat uncertain, appeared to agree well with the estimates from CERN.

The only ingredient still missing in our theoretical picture is the Higgs particle. This may sound incredible bearing in mind that, in some sense, what we call 'empty space' is actually filled to the brim with these things. One could say that they turn the true empty space into a superconductor for the weak currents. These particles are completely invisible as long as they only sit in the lowest energy state; all they do is to give our world its properties as we know them. But it should be possible to disturb them. What then happens can be compared with what the wind can do to a corn field. If we look at the field from some distance, we cannot see the individual stems moving in the wind, but we do see waves of a much bigger size moving across the field. This is an artist's impression that could be given of the *observable* spin 0 particle that the theory predicts, and which we expect the experimentalists to observe in due course.

We can predict many properties of the missing Higgs particle accurately, except for one thing: its mass. This could theoretically be anything between 1000 and about 1 000 000 MeV. By now, the lower values have been excluded experimentally, leaving all values between 60 000 and 1 000 000 MeV still possible at present.† This, unfortunately, is a very uncertain prediction, not of much use to our poor experimentalist friends. Combing this enormously extended region of possible mass values must be done piecemeal.

The mass unit of 1 000 000 MeV, corresponding to the energy an electron or proton would have if it traversed an electric potential of one million million volts, is called a tera-electron-volt (TeV). In the near future, physicists hope to be able to use machines that can reach this amount of energy *per lepton or quark*. For a long time I have suspected that the Higgs particle would be quite heavy, in which case even higher energies will have to be given to these particles before they can create detectable numbers of Higgses,‡ and detection of the Higgses might not come about very soon, but both the very latest data and the latest most sophisticated theories give us more reason for hope; they appear to favor lighter values for the Higgs mass.

This elusive particle *will* be found – and then what? Then we have what we will call henceforth the 'Standard Model'. This is a mathematical description of all known particles and all known forces between them, enabling us to explain all the behavior of these particles. The Standard Model is built exactly according to the recipe of Chapter 12. As far as we know, there is no single physical phenomenon that cannot be regarded as some consequence of the Standard Model, and yet its basic formulae are not terribly complicated. We do admit that the model is not absolutely perfect, and in Chapter 19 we will begin to speculate on how to improve it; however, the degree of perfection reached is quite impressive. Table 7 lists all presently known *fundamental* particles or fields. (Each particle is described by a corresponding field, which is why we sometimes use the word 'particle' and sometimes the word 'field' to denote the same concept: an ingredient in the Standard Model.)

By 'fundamental' we mean to say that, as far as we know, this particle is not built from other particles (or the field is independent of the other fields). Of course, we should not commit the same mistake as before: what may seem to be fundamental now could turn out to be composed of 'even more fundamental' objects tomorrow. So these are the '1995 fundamental' particles. Several aspects of the Standard Model are a bit too complicated to explain here in detail, but I will summarize the most essential parts.

† As of 1995.
‡ To create detectable amounts of 1 TeV Higgs particles, the colliding particles must have considerably more than 1 TeV of energy.

Table 7. *The Standard Model (as it looked in 1995)*

Name	Symbol	Mass (MeV)	Charge	N_c	
Spin 1, gauge photons:					
Photon	γ	0	0	1	$U(1)$
Vector bosons	Z^0	91 188	0	1	
for the weak	W^+	80 280	+	1	$SU(2)$
force	W^-	80 280	−	1	
Gluon	A_s	0^*	0	8	$SU(3)$
Spin 0, Higgs:					
	H^0	$> 60\,000$	0	1	
Spin $\frac{1}{2}$, quarks:					
$I\{$ up	u	5^*	$\frac{2}{3}$	3	
down	d	10^*	$-\frac{1}{3}$	3	
$II\{$ charm	c	1600^*	$\frac{2}{3}$	3	
strange	s	180^*	$-\frac{1}{3}$	3	
$III\{$ top	t	$180\,000^*$	$\frac{2}{3}$	3	
bottom	b	4500^*	$-\frac{1}{3}$	3	
Spin $\frac{1}{2}$, leptons:					
$I\{$ e-neutrino	ν_e	≈ 0	0	1	
electron	e	0.510999	−	1	
$II\{$ μ-neutrino	ν_μ	≈ 0	0	1	
muon	μ	105.6584	−	1	
$III\{$ τ-neutrino	ν_τ	≈ 0	0	1	
tau	τ	1771	−	1	
Spin 2, graviton:					
	g	0	0	1	

N_c is the number of different color components a particle type has. Particles with $N_c = 1$ are colorless. Masses with $*$ superscript are effective masses, which could be defined to be the masses obtained if the surrounding color force fields were ignored, a definition which is not as precise as the other values. Furthermore, evidence is mounting that the neutrino masses are not exactly equal to zero.

We have three types of Yang–Mills gauge fields, each requiring their own 'constant of interaction': these are numbers that describe the strengths of the forces resulting from each of these systems, and they are *not* predicted by the theory but have to be measured by experiment. The mathematical symbols for the forces are $U(1)$, $SU(2)$ and $SU(3)$, and the interaction constants are called g_1, g_2 and g_3. These three constants are all that is needed to describe all forces between the particles. The Yang–Mills fields themselves are associated with particles, the gauge bosons, which have spin 1.

Now we introduce the fermions, particles with spin $\frac{1}{2}$, of which there are three generations. Each generation contains one quark doublet (such as u and d) and one lepton doublet (such as v_e and e). The theory does not prescribe that there have to be three generations; there could have been many more. But when LEP was switched on, the decay rate of the Z^0 could be measured so accurately that the number of neutrino species could be pinned down to exactly three. Had there been more kinds of neutrinos, then the possibility of new decays would have made the lifetime of Z^0 shorter than that observed. In our present construction of the Standard Model, every generation includes one massless neutrino type, so there cannot be more than three such generations – this does not exclude the possibility of *different* kinds of generations in which all neutrinos are heavier than one-half of the Z^0 mass, but we have no indication whatsoever for the existence of such particles.

The particles we have so far all start out being massless. Now we add one particle with spin 0, the Higgs. It is doing all the heavy work. All particles – gauge bosons, fermions, and indeed the Higgs itself – owe their masses to interactions with the Higgs. The Higgs particle itself has never been detected, but its *field* is being felt everywhere. If the Higgs were not there, our model would have so much symmetry that all particles would look alike; there would be too little *differentiation*. For particles to obtain masses, their symmetries must be sufficiently reduced. This has to do with conservation of *helicity*, spin along an axis parallel to its motion, but the details of this are too mathematical to be explained here. The important thing to remember is that all particles owe their masses to interactions with the Higgs field.

Because there are so many ways in which fermions can interact with the Higgs field, not only mass, but also more structure in the theory, arises. This means that the masses will all be different. Also, having as many as three generations, there is even room for interactions that cause PC violation, so that the rare decays of K_{Long} into two pions can also be accounted for in the model.

For the sake of completeness, Table 7 shows one more particle, the so-called *graviton*, believed to transmit the gravitational force. It appears to be an inevitable consequence of the theories of gravity and quantum mechanics, and its spin will have to be 2. It has never been detected, however, and we do not expect to see it in the foreseeable future.

The particle masses, as well as the strengths of their basic interactions, are not predicted by the theory but have to be determined by experiment. This is because these basic parameters of the model are unrelated *constants of Nature*. If we make a complete list of them we find that there are *twenty* such numbers. In any mathematical description of the Standard Model this list of twenty numbers has to be specified. Many of these numbers correspond to the masses in Table 7, others describe other aspects of their mutual interactions, such as the three gauge coupling constants mentioned before. All of these numbers have been measured, some to a great precision. A few others seem to be exactly zero (such as the 'cosmological coupling constant', describing the extent to which *empty space* emits a gravitational field). These have nevertheless been added to the list because there is nothing in the model that forces them to be zero as far as we know.

I would like to stress emphatically how extraordinary the Standard Model really is, even though it contains twenty numbers for which we do not know why they have the values they have, so that we also do not know how to calculate them from first principles. But once these numbers are given, we can, 'in principle', calculate *any other physical phenomenon*. All the properties of the fundamental particles, the hadrons, the atomic nuclei, atoms, molecules, substances, tissues, plants, animals, people, planets, solar systems, galaxies, and perhaps even the entire universe, are direct consequences of the Standard Model. What is more, for most of their general properties, it would not make much difference what the exact values are of those constants that are not yet known very well, such as the Higgs mass. For instance, the effect of the weak force on chemical properties of atoms are extremely difficult to detect (they should give a helical structure to the atoms such that they can distinguish left from right).

I should hasten to add that all these statements are of not much more than philosophical significance and that they mean quite little in practice. We are not at all able to deduce the properties of a cockroach using our Standard Model, and this will never change. Imagine the following question appearing in an examination:

> Calculate the number of segments of *Asellus aquaticus* starting from the Standard Model. You may use the following list for the Higgs mass and CP violating parameters ...

It will never be possible to solve such problems. It is not the intention of theoretical physicists to suggest that they can do the work of biologists or that of members of any other discipline but physics. Our claim is that the forces of Nature responsible for the number of segments of this creature are known, but that the effect is incalculable. We are hardly able to compute the effects of the fundamental forces on a single hadron such as the proton (the results are often more than 50% off target!), so imagine how forbidding the complexity becomes if 10^{22} atoms take the shape of a cockroach! Following the rules of quantum mechanics, it turns out that the properties of *two atoms together* are tremendously more complicated to calculate than the properties of one single atom, and, in turn, that the properties of most atoms are enormously much more difficult to compute than those of the simplest atom, that of hydrogen.

Understanding the forces involved in our Standard Model is important because then we also know which general laws will be strictly obeyed by them. We have the laws of energy conservation, momentum conservation and even *information* conservation; the latter implies that the so-called paranormal phenomena will have to be explained in terms of ordinary physics, ordinary biology, psychology, and so on. You may think about these twilight themes as much as you like, but if no down-to-earth explanations can be given, then they will be in no way compatible with all we know about the Standard Model.

17 Anomalies

Perhaps the foregoing has given you the impression that, in the end, our efforts amount to not much more than continuously discovering new particle types, which are subsequently just added to our model. More or less by accident, we have arrived at a description that, for the time being, fits neatly with all we know at present, and we have dubbed it the 'Standard Model'. But can one now conclude new things from it, something we could not possibly have known in advance, a prediction of something entirely new? The following story about the 'anomalies' could be an answer to this. This chapter is an intermission, a temporary local excursion interrupting our grand journey to the world of the small. We physicists are perfectionists. If our model shows a dirty spot, we want to clean it. History tells us that if we hit upon some obstacle, even if it looks like a pure formality or just a technical complication, it should be carefully scrutinized. Nature might be trying to tell us something, and we should find out what it is.

Some of the mathematical arguments in this chapter are too technical to explain in detail, but most of them can be translated into words. If you agree with many of my friends, and the publisher, that this chapter is unintelligible, skip it. I am going to make an attempt anyway to show to you how shrewd Nature is.

The term 'anomaly' refers to a very special feature in particle theory. It pops up whenever we study the behavior of very light Dirac fermions (particles with spin $\frac{1}{2}$) in an electromagnetic or a Yang–Mills field. The first example was taught to me by Veltman in 1969, when I was still an undergraduate student.

A neutral pion can, in a very short time, transmute into a proton and an antiproton. These usually return immediately into the 'pion state', but they can also annihilate each other, emitting two photons in the process. We indicate this process by a diagram (Figure 19(a)). If we calculate how it proceeds, following the rules known in the 1960s, we find that the pion can decay into two photons this way, and that the time it needs to do this agrees quite well with the $\pi \rightarrow 2\gamma$ decay observed in experiments. This calculation was first carried out by Jack

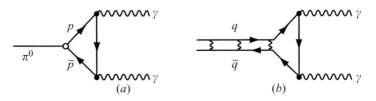

Figure 19. π^0 decaying into two photons: (a) via proton–antiproton annihilation, and (b) via quark–antiquark annihilation.

Steinberger back in 1949. Nowadays, we can perform the calculation more precisely, in a somewhat different manner, because we know how to view the neutral pion as a bound state of a quark and an antiquark (oscillating between $\bar{u}u$ and $\bar{d}d$). It is the quark and the antiquark that annihilate each other while emitting two photons, but the outcome of this calculation gives practically the same outcome as Steinberger's calculation (Figure 19(b)).

So this is beautiful: we understand the neutral pion decay. Or do we? Wait a minute, there is something wrong. The problem is that one can argue that the two-photon decay should have been forbidden, at least to some extent. The 'naked' proton used in the 'proton picture' is practically massless, as is the naked quark in the quark picture. Had they been *exactly* massless, then we could describe a special kind of conservation law inside this pion which would make the decay into any number of photons absolutely impossible. Massless fermions should be unable to annihilate each other here. We say that the 'helicities' do not match.† Now it was already known that this conservation law does not do its job impeccably. The naked quarks do carry a small amount of mass, so that indeed the pion has a chance to decay.‡ But the decay should have been suppressed. The pion lifetime should have been much longer than it was found to be. In short, we had a *formal* argument telling us that there was something not in order.

Not only the experimental data tell us that this 'formal' argument was defective, but also the accurate calculations gave the correct result. How could this happen? Upon analyzing the problem it was realized that the actual calculation had required a 'trick' before producing any sensible answer at all. It had been necessary to *renormalize*. The theory contains infinite forces that cancel each other out when one calculates anything that can be measured. But when one tries to

---

† Helicity stands for the sense of rotation of a particle as compared with its direction of motion, which is well-defined and conserved for a particle moving with the speed of light, that is, 'massless' particles.

‡ A particle with mass flips its helicity with a frequency proportional to this mass.

check what happens to the helicities, it is discovered that their conservation law is sacrificed during the infinity cancellation procedure. This turns out to happen whenever a fermion runs around in a diagram with a triangular trajectory, as in Figure 19. Apparently, our elegant algebraical considerations are being ruined by renormalization. In the case of the pions, theoreticians were satisfied with this answer. The agreement between the accurate calculations and experiment removed any lingering doubt. We call this phenomenon the 'triangle anomaly'.

As far as the pion is concerned the anomaly was not much more than a curiosity, a thing to keep in mind so as to avoid making mistakes. But when I was trying to find the precise prescriptions for the renormalization of Yang–Mills theories, the anomaly showed up again. In the Weinberg–Salam model, the naked fermions have to be *exactly* massless. The very special properties of left and right rotating massless fermions were essential ingredients in this model. The ones with left handed 'helicity' interact with the gauge fields in a way that is different from that of the right handed ones. This is how we could account for the peculiar fact that the weak interactions have a 'corkscrew nature': they distinguish left from right. But the scheme is a delicate one. To prove that all particles involved have acceptable physical properties, we needed to know that all interactions are 'gauge invariant'. Now the anomalies could really cause havoc. Gauge invariance was destroyed: if you calculated something in different ways you would arrive at different answers, and that's bad logic.

Initially, I had hoped that this little bug in the renormalization prescriptions could be 'fixed'. After all, how could one let such a silly thing as an anomaly ruin such a beautiful scheme as the Weinberg–Salam model?

But I was quickly talked out of this.† The anomalies would turn the theory into something useless, *unless* they happen to cancel each other out exactly! Anomalies play a role only in diagrams containing a triangular fermion trajectory. There things could go wrong. But the contributions of all fermion types had to be added together. This left the possibility that they might cancel out, and in that case our theory would be fixed. And now what follows is a little miracle. If we take the Standard Model as it was formulated, all the dangerous anomalies due to the leptons e and v_e on the one hand, and to the u and d quarks on the other, neatly cancel each other out! The same thing happens if one takes the muonic leptons and the s and c quarks together, and finally we have tau, top and bottom. This is why I said in Chapter 16 that we wanted to have exactly as many lepton species as quark species. Perhaps surprisingly to some, Nature

† The honor goes to William A. Bardeen at Princeton.

was sensitive to the theoretician's wishes and gave us equal numbers of leptons and quarks. If this had not been the case, we would have been faced with an impossible renormalization scheme.

The weak interaction theory is not the only one where anomalies play a decisive role. The triangle diagram also occurs in the strong interactions when only gluons and quarks are in the picture. In this case one considers the algebra connected with the conservation of quark helicity (handedness of quark spins in the direction of motion). A consequence of this algebra is that some hadrons are massless and have spin 0: these are the Goldstone bosons mentioned in Chapter 11. Indeed, the three pions have very little mass,† and their spin is 0. However, the rules of our mathematics told us that there should be four conservation laws rather than three, and that therefore there had to be a fourth light, spin 0 particle. The only object that would meet this and some other necessary specifications was the èta particle, η. In short, quantum chromodynamics predicted that, besides the pions, the η particle had to be comparatively light.

Yet the èta is much heavier than the pion. The problem was aggravated by the fact that it is the *squares* of the masses that have to be compared. This was called the èta problem. Quantum chromodynamics seemed to predict a light èta particle, and it was not there. Now it was realized that, yes, there is an anomaly when we want to formulate the argument for the èta particle accurately. Again, one could imagine that the èta could temporarily turn into two gluons, where, just as in the case of the massless pion, triangular diagrams play a role. This time, however, accurate calculations could not be performed, and it was very difficult to imagine how Nature could manage to turn the anomaly into mass for the èta particle. The mathematical rules had to be violated, yet the helicity conservation seemed to be obvious. This was because when one applied the argument for this case, a subtlety was being overlooked. The true origin of the anomaly is something much deeper than the renormalization procedure; it is in the nature of the distinctions we make between a *particle* and an *antiparticle*.

A very 'primitive' description of the fermions (the way Dirac used to formulate the theory when he was the first to come to grips with the situation) is one in which we start with no such things as antiparticles at all. Instead, what we have are 'locations in space', each of which may or may not be occupied by a particle.

---

† That the pion mass is not exactly zero is then attributed to the fact that the quarks have a tiny, non-vanishing 'naked' mass.

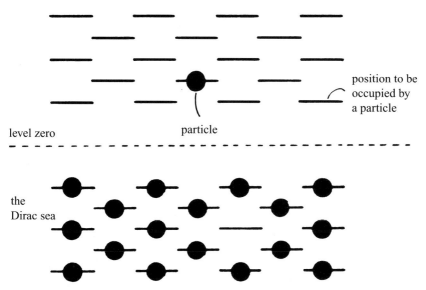

Figure 20. In this figure, there is one particle and one antiparticle present, according to Dirac's theory.

Suppose we make a list of these locations and the energies they represent.† If for one such location this energy is negative, we put a particle in it; if the energy is positive we keep it empty. If our entire stock of particles precisely matches the need to do this for all locations, we end up with the 'state with lowest possible energy' (as will soon become clear). We call this state *empty space* or *the vacuum.*

But perhaps we have one particle left over. This particle will then have to be put into a location with positive energy, and then it is a real observable particle. Perhaps we fall one particle short. This means that we have to leave a location with *negative* energy *empty.* To make such a 'hole' with *negative* energy actually requires a *positive* amount of energy. Such a hole behaves also exactly as if it were an ordinary particle, with positive energy, except that its electric charge (and other properties) are opposite to that of the original particle. So this is what we call an *antiparticle.*

Now we can explain what an anomaly is. Under certain, special circumstances, when strong fields are present, a location of negative energy can turn into one of

---

† This list is infinite, so we have to stop once the energies are too far away from zero. This is why the way in which infinities are suppressed in the theory, called 'renormalization', comes into the picture.

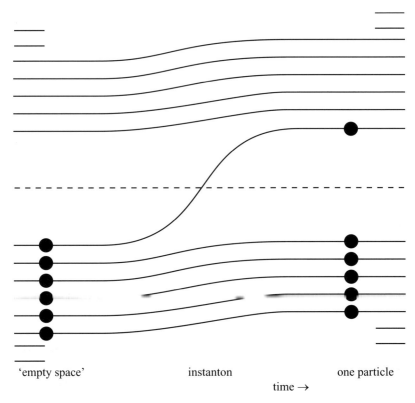

'empty space'   instanton   one particle

time →

Figure 21. An instanton can cause a transmutation of a negative energy position into a positive energy position (or vice versa), dragging its occupant along.

positive energy. Suppose this location was originally occupied, meaning that it merely comprised the empty space. When its energy turns into a positive value, suddenly, like a *deus ex machina*, a particle appears out of nowhere. If the original location was empty, then this phenomenon would appear to cause the complete disappearance of an antiparticle. A positive energy location can also turn into a negative energy one, and then the opposite process takes place.

The èta problem came about because this description of particles and antiparticles, as Dirac had seen them, had been more or less forgotten, but, most of all, because people had not realized that such transitions from negative to positive energy and back were really possible. While toying with the corresponding equations, four Russian investigators, Alexander Belavin, Alexander Polyakov, Albert Schwartz and Yuri Tyupkin, discovered a gauge field configuration that looked like a 'twisted knot' in space-time. It was realized a little later that this 'knot' was precisely the entanglement of field lines necessary to cause a negative–positive energy transition for a fermionic state. The field configuration looked like some 'particle', but one that is present only during an instant of time, disappearing as quickly as it was formed. So it was given a name that emphasizes this: an 'instanton'. An instanton is like a little door that suddenly appears, opens to let one or several particles through, to or from this infinite reservoir called the 'Dirac sea', and then closes and disappears.

And now we know that it is instantons to which the èta particle owes its mass.† What an instanton does in quantum chromodynamics is the following: it turns the energy of one right helicity particle state from positive to negative and does the opposite to one left helicity state. So one right helicity particle will seem to disappear and a left helicity particle pops up. It is as if a right helicity particle transmuted into a left helicity one! This is why helicity is no longer conserved, and consequently the algebra that was associated with it breaks down. The èta particle really is as it should be in QCD.

It is not allowable to have a theory in which particles that are not conserved interact with a Yang–Mills field, for the same reason why our theory of the electromagnetic forces only works when electric charges are exactly conserved. Therefore, because of the behavior of instantons, left helicities are not allowed to interact with gauge fields independently from right helicities. If this impossible thing is attempted, we run into a problem – the problem we called an 'anomaly'! It seems that more and more details of our theory are coming together.

Until now, perhaps it could be maintained that all we are doing is fitting observed phenomena into our model, although I can assure you that the anomalies could never have been argued away differently. Now comes the new thing. The instantons we discussed are the ones for the strong $SU(3)$ gauge fields called quantum chromodynamics. But they must also occur as weak $SU(2)$ gauge field configurations. The 'weak' anomaly, the one that forced us to have exactly as many lepton as quark generations, could have a more direct effect. The weak interactions *do* act differently on the different helicities; they do distinguish left from right. Therefore, the instanton does dangerous things here. If there were just one generation of quarks it could remove one of each color: three quarks would disappear. If there were one generation of leptons, one lepton would disappear too, or an antilepton would come out, which amounts to the same

---

† I should add that this explanation of the èta mass is still being disputed by some, who claim that the instanton cannot completely remove the symmetry that protected the èta from becoming massive. I insist, however, that, provided one formulates things sufficiently carefully, the instanton argument is entirely correct.

thing. Now, since there are three generations, all feeling the same weak fields, a single instanton does all these things to each generation simultaneously. What results is an interaction of the type

$$u + u + d + s + c + c + t + t + b \rightarrow e^+ + \mu^+ + \tau^+ \,.$$

If you look carefully, you will notice that total electric charge is exactly conserved in this transition. This would not have been the case if we had omitted the τ lepton (and its neutrino); the interaction would not have been allowed. You cannot have electromagnetism without charge conservation. So this is why a theory without the τ lepton pair would have been faulty.

Since baryons each consist of three quarks, and the exotic baryons can turn into ordinary baryons by means of 'ordinary' weak interactions, *three* baryons (for example three protons) can disappear and leave three positively charged leptons (or antineutrinos and pions) in their place. Eventually, all matter in the universe (which, after all, consists mainly of baryons and leptons) could decay into much lighter particles and finally annihilate itself completely! A spectacular result, except when we perform the corresponding calculation more precisely. The indicated transition will be *extremely* rare. During the very early stages of the universe, there probably was a short period during which interactions of this type were very frequent. Perhaps we owe the present surplus of baryons and leptons in the universe entirely to the reverse of this process! Unfortunately, our understanding of the first stages of the universe is so rudimentary that it is not (yet?) possible to elaborate on the details of such ideas – although several investigators are trying just that!

What is remarkable in this anomalous event is that we started with a theory with absolute conservation of baryon number. Nevertheless, we have *derived* that baryons are not immortal. Later, we will encounter other theories where baryons decay in a different way. This means that all varieties of matter will ultimately be unstable, but I can reassure you: many years will go by before a single proton in your body will decay (and even much more time before three will decay together), and even when this happens, you won't feel a thing.

18 Deceptive perfection

So here we are. Apart from just a few minor technical details, theoretical physics is finished. We have a model that encapsulates everything we wish to know about our physical world. What else do we want?

Well, the Standard Model is nearly, but not quite, perfect. First of all, we could begin to complain about those twenty uncalculable constants. But if this were the only complaint there would probably be little we could do about it. Of course, numerous ideas have been suggested to explain the origin of these numbers, and theories have been put forward that purportedly 'predict' their values. The problem with all these theories is that the arguments they give are never compelling. Why should Nature care about some magic formula if without such a formula no real contradictions arise? What we really need is some fundamental new principle, such as the relativity principle, but we do *not* want to abandon all the other principles that are already known; these, after all, have been so tremendously helpful in discovering the Standard Model! The best place to look for a new principle is where our present theory has other weak spots.

A universal rule in particle physics is that, when particles collide with greater and greater energies, the effects of the collisions are determined by tinier structures in space and time. Suppose we had at our disposal a particle accelerator in which particles can be given 1000 times the energy that can presently be reached. The collisions that would take place would tell us something about structural details inside those particles, that are much smaller than before. Will the Standard Model still be correct there? Let us, like Gulliver, continue our journey towards a world of still smaller objects.

The Standard Model is a mathematical construction. It predicts unambiguously what the world of still smaller structures should look like. Alas, there are several reasons to suspect that such predictions will turn out to be completely false.

Let us switch on our imaginary super-microscope, and focus it right to the center of a proton, or any other particle. We see hordes of naked fundamental particles frolicking about. As seen through the super-microscope, the Standard

Model contains twenty constants of Nature, describing the forces that rule the way they move. However, these forces are now not only quite strong, but also they cancel each other out in a very special manner; they are *tuned* to conspire in such a way that the particles behave as ordinary particles when you switch the microscope back to ordinary magnification scales. If, in our mathematical equations, any one of these constants were replaced by a slightly different number, then most of our particles would immediately obtain masses comparable to the huge energies that are relevant in the very high energy domain. The fact that all particles have masses corresponding to much lower energies suddenly becomes quite unnatural.

This we call the *fine-tuning problem*. As seen through the microscope, the constants of Nature seem to be very carefully tuned, for no apparent reason other than to make the particles look the way they are. There is something very wrong here. From a mathematical point of view, there is no reason to object, but the credibility of the Standard Model plummets if you look at extremely tiny distance and time scales, or, and it amounts to the same thing, if we calculate what would happen if the particles collide with extremely high energies. And why *should* the model still be valid there? All sorts of super-heavy particle types could exist that can only be brought to life at energies as yet unattainable, and they could modify completely the world our Gulliver was planning to visit.

In short, if we wish to avoid the need of delicately fine-tuned constants of Nature, we create a new problem: how can we modify the Standard Model in such a way that fine tuning is no longer needed? That modifications are then necessary is certain. It implies that very probably there is a limit beyond which the model as it stands ceases to be valid. The Standard Model will be nothing but a mathematical approximation that we have been able to create such that all *presently observed* phenomena are in agreement with it, but every time we switch on a more powerful machine, we must expect that modifications will be needed.

How could we ever have thought otherwise? Whence this 'arrogance' to think that we might have the 'ultimate' theory? On looking at things in this way, our present problem may well be the opposite of the question of where the Standard Model ends: how could it be that the Standard Model works so extraordinarily well, and why have we still not been able to perceive anything like a next generation of particles and forces that do not fit the Standard Model?

The question 'What is there beyond the Standard Model?' has been haunting physicists for years now. Up to this point, I have been able to report to you with some authority on all those things that we know. From now on, I am going to speculate about what we do not know. Of course, one can always imagine that physics as we know it will entirely cease to be valid and be replaced by something completely different. But this we do not believe. If history has taught us one thing it is that, with hindsight, newly discovered laws always turn out to be quite logical extensions of what we have already known for a long time.

19 Weighing neutrinos

There are two directions in which one can try to extend the presently known Standard Model, and these can essentially be characterized as follows:

(1) rare new particles and extremely weak new forces, and
(2) heavy new particles and new structures at very high energies.

Let us begin with the first. Particles could exist that are very difficult to produce and to detect, and, for that reason, they could have escaped our attention up till now. All particle types that exert strong forces on any of the known particles could never hide themselves from us. According to the forbidding laws of quantum particle theory, such particles would be frequently produced, either singly or in particle–antiparticle pairs – this would only not happen if their masses are too great, but that is covered by case (2). They would betray their presence by colliding against particles in a detector. The only light particles that can avoid our detection are ones that exert only very tiny forces upon most or all known species.

The first additional particle that springs to mind is a *neutrino spinning to the right*. Recall from Chapter 7 that neutrinos only rotate to the left (antineutrinos rotate to the right), if you take the axis of rotation parallel to the direction of movement. From a mathematical point of view, there is no objection to this, but it is somewhat unaesthetic. After all, in each particle generation there are, besides leptons, also quarks, and all these can rotate to the right. We say that the quarks, as well as the charged leptons, have right helicity components. Suppose now that neutrinos also possess right-handed components. Since rotational motion is then no longer coupled to linear movements, neutrinos in this case may have a mass (the mathematical reason for this relationship between spin and mass, which I mentioned earlier, is not very easy to explain).

Neutrinos have always manifested themselves as if they were strictly massless. It looks as if they move exactly with the speed of light. But there is a limit to the precision of our measurements. If neutrinos were very light, for instance just one-hundred-thousandth of the electron mass, we would be unable to detect the difference between these and strictly massless neutrinos in the laboratory. But as I said, for this a right-handed neutrino component is required.

At this point, astronomers join in the discussion. It is not the first, and it will not be the last, time that astronomy can provide essential information concerning the elementary particles. For example, because of *neutral current interactions* (the weak forces originating from Z^0 exchange), neutrinos are a crucial factor in the *supernova* explosion of a star. When nothing was known about the neutral current, it was thought that neutrinos produced during the explosion would be able to escape from the star unhindered. Now, however, we know that due to neutral current interactions they can collide against the outer layers of a star and blow these away with tremendous force. This new supernova theory strongly supported the neutral current hypothesis, long before the Z^0 had been discovered.

For about twenty years, astronomers have been pointing out a small but stubborn discrepancy in their 'standard model': the theory of the interior of the Sun. The Sun is a giant nuclear reactor. Because it is so close to us (compared with other stars), extensive measurements are possible. These measurements yield the chemical composition, mass and temperature of the Sun. Many of the nuclear reactions taking place inside the Sun can be mimicked in the laboratory. Others can be calculated with high precision. Only a few of the reactions are less accurately known. One can deduce the composition of the materials from which the Sun was made some six billion years ago, for instance, from measurements on meteorites. And so a model could be constructed of our Sun.

This solar model works fine, and most of the other stars in the universe can be described using the same techniques. The solar model tells us clearly and precisely what the temperature and density are deep in the interior of the Sun. A fantastic way to check the model is by registering vibrations in the Sun, or 'sunquakes'. Tiny variations in the diameter of the Sun are found, which indicate that the surface vibrates. Astronomers then do what seismologists do with earthquakes: from their measurements they deduce properties such as density, composition and temperature in the solar interior.

All these findings agree beautifully, we are assured by astronomers. However, one thing does not agree with the rest of the theory. Various nuclear reactions going on in the solar interior should also produce neutrinos. These should mostly be neutrinos of the electron type (v_e and $\bar{v}_e$) because the others could only arise together with muons and taus, and for those the energies of the particles in the Sun are not sufficient. Several experiments have now been carried out that are aimed at detecting these solar neutrinos. These measurements were difficult because of the extreme inertness of neutrinos. Finally, after a great deal of effort, they were successful, but the neutrino flux from the Sun was found to be quite a bit less than anticipated. Many attempts were made to understand this discrepancy and to sweep it under the carpet. For instance, the neutrino production depends very strongly on temperature. Could it be that the solar temperature is lower than expected? Whatever improvements one tried in the calculations, this was not what came out: the temperature had to be right. Also, a lower temperature would also not agree with the solar vibration measurements. The experiments were improved, and new methods were invented to measure the neutrinos. The investigators now all agree about the disagreement: there are at least three times as few neutrinos as there should be. Therefore something is wrong with our models.

What does this have to do with the neutrino mass? Well, a neutrino with mass could make transitions into other neutrino types. The situation is comparable with the description I gave of the K-mesons, but is somewhat more complex. If v_e's could turn into v_μ's or v_τ's, they would not leave a trace in our detection apparatus, so that it would *seem* that there are fewer neutrinos. We call this 'neutrino oscillation'. Neutrino oscillation is only possible if neutrinos are not massless as in the Standard Model.

For a long time I found it difficult to believe the neutrino oscillation hypothesis as an explanation for the missing solar neutrinos. I had thought that only a few neutrinos could really escape along that route. After all, they should be able to rotate back into v_e's exactly as easily. But in 1985 a very ingenious theory was put forward by the Russians S. Mikheyev and A. Smirnov, which was based on an earlier idea of PC-expert Lincoln Wolfenstein. The so-called MSW mechanism works as follows.

As the solar neutrinos oscillate, they also interact, though extremely feebly, with matter in the Sun. Due to the presence of electrons in the Sun, neutrinos of the electron type can make transitions into W bosons and back, whereas the other neutrino types can only transfer via the Z^0 bosons. A consequence of this is that the velocity of electron-neutrinos inside the Sun difers very slightly from that of the other neutrino types. It was calculated that by the *combined effect* of these interactions and the neutrino oscillations, electron-type neutrinos could nearly completely metamorphose into, for instance, muon- or tau-neutrinos.

This argument surely consists of a long chain of calculations and hypotheses, and it is only as strong as its weakest link, but it seems to amount to a serious indication that neutrinos have a mass (somewhere between 0.005 and 0.5 eV, which is less than one-millionth of the mass of an electron). So the zeroes in Table 7 are probably wrong!

Right-handed neutrinos are insensitive even to the weak force. Only with the help of the Higgs particle can they first turn into left-handed neutrinos, after which they can undergo (rare) weak interactions of the conventional type. How many more of such hidden objects could exist?

I have already mentioned the *graviton*. This hypothetical particle feels only the gravitational force (of which it is the carrier), and this gravitational force is much weaker than the weak force. The fact that we experience the gravitational force every day is merely a consequence of one property of the gravitational force that distinguishes it from all other forces: all the atoms contained in the Earth are pulling at all the atoms inside our bodies *in the same direction* (towards the Earth). Because there are so many atoms in our bodies, we notice this force. But imagine being in a spaceship in a free-fall trajectory, where the pull of the Earth cannot be felt. The spaceship also has many atoms pulling at bodies inside it in this or that direction, but not nearly as many as in the entire Earth. This is why astronauts in a spaceship notice hardly any gravitational force at all (it can be detected, but only with very special equipment).

The gravitational force acts over enormous distances, and this is directly related to the fact that its carrier, the graviton, has a rest mass equal to zero. One could ask the question whether there are more such extremely weak, but far reaching forces. For example, one could think of an extra component in the gravitational field that only acts on electrons, or only on baryons, or on one of the exotic quark types, so that it is different from ordinary gravity, which is known to act on all of these, discriminating between the particles exclusively on the basis of their masses and nothing else. The Hungarian baron Roland von Eötvös carried out careful and ingenious experiments at the end of the nineteenth century to see if there are any such non-standard selection effects in gravity. More than half a century later, these measurements could be improved by Robert Dicke (who, in spite of having access to much more modern electronics, had quite a lot of difficulty being more accurate than Eötvös). Searches were also carried out for deviations from the gravitational $1/r^2$ law, which would occur, for instance, if there existed a weaker component in gravity that acts only over a limited range of distances (which would imply that the carrier of this new field would have a rest mass that is just slightly bigger than zero).

With these particular possibilities in mind, Eötvös' own notes of his experiments were recently subjected to re-examination. In the 1980s, it was claimed that the notes indicated a slight bias in the gravitational force, namely that different kinds of materials would react slightly differently upon a gravitational field. Eötvös himself had always maintained that the minute irregularities in his data were insignificant, and current thinking is that he was probably right. However, a name for the purported new force was quickly found: the 'fifth force'. In spite of elaborate attempts, it has never been possible to confirm the existence of such a force. Theoretically, the existence of a new force resembling, but different from, the gravitational force, cannot be excluded, but in my opinion adding a force of this sort to the Standard Model would make it quite a bit uglier than it is now. I am not giving such a force much of a chance.

Finally, one could think of many other species of very weakly interacting particles. Do they exist? Remember that up till now only three species of neutrinos have been found. Other neutrino species very probably do *not* exist. This we know from accurate LEP experiments. If there had been more neutrino species than v_e, v_μ and v_τ, then the Z^0 particle would also have been able to decay into those, and this would have shortened the Z^0 lifetime by an observable amount. The measured lifetime corresponds exactly to there being three neutrino types.

If the elementary particle world contains a large number of extremely weakly interacting particle types, instead of only three, which furthermore have a very definite role to play in the Standard Model – the three neutrinos, then why do we not see any particle at all that interacts 'a little bit stronger than extremely weakly'? Personally, I think it is because there exist only very few such ultra-weakly interacting particles, if any at all.

True, I do not carry the stone of wisdom in my pocket, and my 'educated guess' may be completely wrong. Quite a few extensions of the Standard Model have been invented in which there is indeed a place for ultra-weak particles. And, here again, astronomers may have something to tell us. They claim that outer space contains a form of matter that does not interact in an ordinary way with stars, gas and dust clouds, but betrays its existence exclusively by interacting gravitationally with stars and galaxies. What kinds of particles comprise this 'dark matter' nobody knows as yet. Could they be perhaps unknown heavier brothers and cousins of neutrinos, too heavy for the Z^0 to decay into, the so-called WIMPs (weakly interacting massive particles)? Ideas and theories abound (there are nearly as many as there are physicists who have any opinion about them).

20 The Great Desert

Gulliver boldly continues his journey. When protons and neutrons are magnified 1000 times, the Standard Model tells him what details as small as 1/1000 of their diameter will look like. What comes after that, at even greater magnifications, is uncertain. We are now entering the world of very *heavy* particles, carriers of forces over ultra-short distances. We will concentrate on structures at distance scales between 1/1000 and 1/10 000 000 000 000 000 000 (that is, 10^{-19}) times the proton diameter.

This seemingly absurd figure implies that the territory we are entering now covers some sixteen orders of magnitude (sixteen more zeros). This is about as much as the difference in size between a house and an atomic nucleus.

Surely this new world could be just as complicated as the previous one, the one described in the preceding nineteen chapters. But it could also be quite a bit simpler, in the sense that it might be possible to extrapolate all of the laws of physics over the whole area. It certainly seems as if the rules which I described in Chapter 11 will remain irreplaceable. Perhaps we will continue to find other particles and other fields, but no matter how I adjust the magnification of my imaginary microscope, I will see the same ground rules for objects with spin 1, spin $\frac{1}{2}$ and spin 0 (as long as the *gravitational* force may be neglected, but more about that later).

There are hardly any experimental data known about this world, so we cannot say much about it with certainty. Nevertheless, it is generally assumed that the ground rules will be valid. As for the rest, we can call upon our imagination. Theorists then come up with three kinds of 'scenarios'. We begin with what Raoul Gatto in one conference called the 'zeroth scenario'.

According to the zeroth scenario the Standard Model is entirely correct. There is no further structure. The laws of physics in our new world will have to be determined by phenomena happening at the still smaller scales of 10^{-19} proton diameters, and for all distances larger than that the Standard Model applies. In this case, there is a terrible fine-tuning problem: at 10^{-19} proton diameters, the buttons of the constants of Nature have been adjusted extremely delicately. It is as if someone puts a pencil upright on a table, on its tip, in such a way that it falls over after 19 minutes. How this could be achieved will remain a complete mystery to us, which we simply ignore; let it be a problem for the philosophers.

From a mathematical point of view, this is just fine. There will simply be nothing in this enormous area of physics. We speak of it as the 'Great Desert'. By a fluke of history our present generation of physicists 'guessed' what Nature is like in a world about as large as that of all preceding physics taken together. We guessed that there is 'nothing'. Whoever believes that this guess is correct, please raise their hands.

No, most physicists think of other scenarios; they think of all sorts of flowers blossoming in the desert. But what will these be like? The only lead we have is the fine-tuning problem. Could we dream up a scenario, a *possible* model, perhaps only describing a part of the desert, but in such a way that nobody had to adjust the buttons with unreasonable precision?

The easiest way to do this is by searching for a *symmetry*, such that the buttons are adjusted in a symmetric way. The symmetry most suitable for this is called 'supersymmetry'. Bruno Zumino, Julius Wess, Peter van Nieuwenhuizen, Sergio Ferrara and many others are the pioneers in this direction. What is supersymmetry?

We saw that, in the Standard Model (but also in its precursors), particle species have been arranged in multiplets. All the particles in one multiplet always have the same spin. No big wonder: the 'manuals' for spin 1, spin $\frac{1}{2}$ and spin 0 seem to be all very different. Symmetry relations between different spins seem to be difficult to imagine. In particular, particles with half-integer spin, the fermions, obey Pauli's exclusion principle, whereas the ones with integer spin, the bosons, show collective behavior (they like to gather together in the same state of position or movement). It is therefore quite astonishing that a mathematical scheme was discovered that put fermions and bosons into the same multiplet. The differences between fermions and bosons are in our description of them; in this proposed scheme they are identical.

The rules for these 'super-multiplets' are very strict. In one super-multiplet one always has exactly as many fermions as bosons. The original (naked) masses of all particles in a super-multiplet are the same. The simplest case we call $N = 1$ supersymmetry. There, all members of a super-multiplet may only differ by one half unit of spin. We also have $N = 2$ and $N = 4$ supersymmetry, featuring larger super-multiplets. It would lead us too far to explain exactly what N means here (but roughly it is the number of half spin steps in a super-multiplet).

In Nature (and, in particular, in the Standard Model) the masses of the fermions are not anything like those of the bosons. Worse still, it seems that there is not a single fermion in the Standard Model that fits with any boson in it to form one super-multiplet. If there were just a shred of truth in supersymmetry, we would have to find 'super-partners' for all particles known to us today. Apparently all these objects will have masses so large that they could not yet have been produced in our machines.

The photon, the W, the Z, the gluon and the Higgs should all have super-partners with spin $\frac{1}{2}$. We call these super-partners 'photino', 'Wino', 'Zino', 'gluino' and 'Higgsino', respectively, names inspired by the word 'neutrino' (even though they have little to do with that; the neutron and the neutrino cannot be super-partners). The super-partners of the quarks and leptons ('squarks' and 'sleptons') must have integer spins.

The mass differences between super-partners in one super-multiplet must be at least millions of MeV's. This certainly means that supersymmetry cannot be exactly right but must be 'broken'. The extent to which supersymmetry is broken may seem to be very large to us, but in the eyes of a Gulliver walking around in a world where many billions of MeV's still seem to be small, this breaking may perhaps be negligible. So it could be that there supersymmetry works with a great precision.

Of course, in such a theory, we would also have to indicate why and how supersymmetry is broken. It is very difficult to invent a credible mechanism for this. It is tempting to think of some kind of condensation process, analogous to the Higgs mechanism, which after all also gave us the masses of the W and the Z, and which was also responsible for the mass differences between the electron and its neutrino, or the muon and its neutrino.

The Higgs mechanism is responsible for all masses in the Standard Model. So why could it not also produce the mass differences in the super-multiplets? As I elucidated earlier in this book, a theory should not consist only of words, but also of accurate mathematical rules for calculations. It turns out that the idea of 'spontaneous supersymmetry breaking' is not good enough to explain the disturbances in supersymmetries. For instance, one would expect a strictly massless particle with spin $\frac{1}{2}$, the 'Goldstino'. Could that perhaps be one of the neutrinos? Daniel Freedman and my fellow student, now colleague, Bernard de Wit showed that the known properties of the neutrino are incompatible with this. It has to be done in a different way, and exactly how we do not know. This is a somewhat delicate issue in the supersymmetry theory.

Once we have (approximate) supersymmetry, we may have doubled the number of particle species in Nature, but what we have gained is the resolution of the fine-tuning problem. Just a few flowers in the big desert, and in addition a super-highway leading to the other side.

Another peculiarity in supersymmetry theory is that the transmitter of the gravitational force, the graviton, must also have a super-partner, the 'gravitino'. The gravitino is then the only fundamental particle with spin $\frac{3}{2}$ (the Delta resonance, Δ, also has spin $\frac{3}{2}$, but it is composed of quarks which have spin $\frac{1}{2}$). This should worry us: the graviton and the gravitino violate our basic rules for renormalizable theories, and so they will ultimately overthrow the entire picture. The grave consequences of this will be discussed later, but as long as these objects only interact very weakly, we are safe. Also to be discussed later is the nature of the theory obtained if one combines supersymmetry with gravity.

21 Technicolor

Supersymmetry has beautiful mathematics, and so the professional literature is full of it. As we experienced earlier, for instance when the Yang–Mills theory was proposed, we have a brilliant mathematical scheme, which we do not yet know how to fit into the system of existing laws of Nature. It does not make sense, as yet, but we may hope that it will do some time in the future.

There is another scenario, actually much more appealing to our imagination. We have seen that atoms consist of smaller constituents, the protons, neutrons and electrons. And then we discovered that these constituents, in turn, have a further substructure: they are built from quarks and gluons. Why, as you might already have thought earlier, do not things go on like that? Perhaps these quarks and gluons, and also the electrons and all other particles still called 'elementary' in the Standard Model, are in turn built out of yet smaller grains of matter?

You would not be the first to have this idea. I have already reported how Jonathan Swift pictured the world of the small as a carbon copy of the world of larger things. Big fleas carry little fleas on their skins, and so on, *ad infinitum*. Well, just as biologists would try to explain to you that the kingdom of the fleas has to be looked upon somewhat differently, I must also state that the picture of an infinite repetition of building blocks cannot be correct as such.

Let us look at the quarks in a proton. Quantum mechanics, the marvellous theory controlling everything in the micro-world with incredible precision, requires that the product of mass and velocity, called 'momentum', must be inversely proportional to the sizes of the 'box' in which you put your system. The proton can be regarded as such a box. It is so small that the quarks in there will have to dash around with near light velocity. Because of this, the effective masses of the two lightest quarks, u and d, will be much bigger than the values listed in Table 7, or something around 300 MeV; this explains why the proton mass is around 900 MeV, which is much bigger than the masses of the quarks (and gluons) when at rest added up together.

In contrast to the proton, the quarks themselves, and also the leptons and all other particles in the Standard Model, seem to be 'point-like'. By this we mean that their properties do not change even if we put them in a box a thousand times smaller than a proton. Now here is our difficulty: suppose that they were composed of other, smaller constituents. These would then surely be packed extremely tightly, and therefore they would have to have much more kinetic energy (energy due to their rapid movements), and they would add correspondingly to the total mass of the thing they are in. Ergo, why are quarks and electrons so light?

This can be said in a more complicated way. Quarks inside a proton have three kinds of 'mass'. First we have what we call the 'free mass', or the mass the object would have if you isolate it. Well, to isolate a quark out of a proton you need an infinite amount of energy, and therefore the free mass of a quark is infinite. This is a meaningless, and therefore useless, concept. Secondly, we have the effective mass of a quark in a proton, due to it being forced to move around rapidly by the laws of quantum mechanics. This so-called 'constituent mass' is about one-third of the proton mass, or 300 MeV. The third notion of mass is the 'algebraic' mass. This is a parameter determining the proportion of this object called the 'mass term' in its equations. For other particles, this mass term corresponds to their real mass; for the u and d quarks, this quantity is only around 10 MeV. Our problem is that our hypothetical new building blocks would have to have terribly large constituent masses, certainly many times larger than the mass of the object they form. It is as if you were asked to build a featherlight racing bicycle out of rock-solid beams of metal.

Yet there is hope: Nature itself has given us an example of how something like this can be achieved. The pion also consists of quarks. The pion is not much bigger than the proton, and so you would expect the quarks in there also to have constituent masses of around 300 MeV. Instead of 600 MeV, however, the pion weighs a mere 135 MeV. This is because the pion mass is protected by a symmetry: the pion is an (approximate) Goldstone boson (see Chapter 12).

This means that maybe there is a way to view particles which themselves are as light as an electron yet are built out of 'heavier' building blocks. You must introduce symmetries, perhaps as many as there are particles in the Standard Model. Then you can explain that all particles we know presently are as light as they are because their masses are 'protected' by a symmetry. It turns out to be a difficult task to elevate this idea into an accurate mathematical prescription.

What has been tried extensively is a repetition of the theme of 'color confinement'. The color forces were so cunningly efficient in keeping the quarks together in the proton and the pion. Maybe one can construct a new version of such a color theory, at a scale one thousand times smaller than the old color theory, keeping constituents together in what presently is called elementary particles. There are versions of such constructions that do not seem to be altogether impossible. Theories of this sort were dubbed 'technicolor': a color one thousand times more powerful than that of quantum chromodynamics.

I once proposed to view the quarks as the 'fourth building blocks' (molecules consist of atoms; atoms of subnuclear particles; subnuclear particles of quarks). If quarks in turn consist of building blocks, for the fifth time, then an appropriate name for these would be 'quinks'. This leaves open the option later to introduce 'sexks', 'septemks', and so on. I am not so fond of using superlatives such as 'super' and 'hyper' to describe new things (think of 'supernova', 'superconductivity', 'supersymmetry' and so on). We cannot go on using such bombastic language forever. Serious investigators who do so remind me of the creations of the well-known Dutch cartoonist Marten Toonder, two of whom call themselves 'Super' and 'Hieper'. They are crooks. Unfortunately, my proposal to name the new objects 'quinks' has not been adopted. One more often reads 'preons'. Apparently 'pre' here means that they must come before the 'protons', which already came first. People have not yet learned.

There are quite a few difficulties in the technicolor theory. The electron and the muon, for instance, are as alike as two drops of water. It would have been natural, and economical, if they were built out of the same building blocks. But then it should be possible for the muon to turn into an electron while emitting a photon. The hadrons in the ordinary color theory make exactly such transitions in a way that agrees with calculations. But a muon never decays into an electron and a photon. There are always neutrinos involved in the game.

The most important thing a technicolor theory could do is to find the building blocks of the Higgs particle, the particle that presents the biggest problems with fine tuning. Also, we like being encouraged by precedents in theoretical physics. There are at least two cases where a composite particle caused spontaneous symmetry breaking: one is the old sigma model of Gell-Mann and Lévy (the sigma being actually composed of quarks), and the other is the BCS theory of superconductivity, where a phenomenon similar to the Higgs mechanism is caused by the bound state of two electrons (Cooper pairs, see Chapter 11). But in the Standard Model many of the properties of the Higgs particle are precisely known because they are responsible for the masses of known particles. Our theory should be able to 'predict' these interactions, and, alas, that does not lead to even a remote agreement. New gauge theories are needed to generate interactions responsible for the masses of the known particles, and so we end up with 'extended technicolor theory'. And then there is a danger of generating too many new kinds of interactions. Several versions were disqualified because they would disturb the delicate equilibrium between the neutral kaons that was responsible for their unique properties (see Chapter 7).

I have found rules one has to obey when a symmetry pattern is designed for a technicolor theory, in which I made use of the anomalies mentioned in Chapter 17. These rules contributed to the demise of the entire idea. It seems to be nigh impossible to obey these rules and still obtain a credible model. Yet, it could be that the real world is governed by a scheme of technicolor forces that is so complicated that we could not possibly have guessed it from what we know now. Perhaps new accelerators such as the 'Large Hadron Collider' (LHC), to be completed shortly after the turn of the twenty-first century, will unveil some of the new phenomena associated with such schemes. That the United States government decided (after expenditures of billions of dollars) to halt the construction of their 'Superconducting Supercollider' (no comment about the name) was a serious blow to high energy physics; many physicists had put all their hopes on information to be gathered from this machine.

There are, by the way, also attempts to combine ideas from supersymmetry with those from technicolor. It should interest astrophysicists that such theories not only predict multitudes of super-heavy new particles, but also several types of ultra-weakly interacting particles, the 'techni-pions'. They may be the WIMPs populating the voids between the galaxies, thus being responsible for the missing mass for which astrophysicists are still searching.

22 Grand unification

When in 1931 Paul Dirac concluded from his equation for the electron that there should exist an antiparticle with opposite electric charge, he felt very embarrassed. Such a particle had not yet been discovered, and he really did not want to disturb the scientific community with such a revolutionary proposition. 'Maybe this strange positively charged particle is simply the proton,' he suggested. When shortly after that the real antiparticle of the electron (the positron) was identified, he was so surprised that he exclaimed: 'My equation is smarter than its inventor!'

Nowadays, physicists no longer suffer from such modesty. High energy physicists are now accused of arrogance, and not fully without reason: when the new gauge theory for the weak force became popular, Weinberg and Salam had no trouble at all advertising it as a 'unification' of the weak force with the electromagnetic force; and to avoid any misunderstanding: 'the most important unification since Sir James Clerk Maxwell unified electricity with magnetism'.

I have always maintained that the Weinberg–Salam model does not fully deserve such laudable comment. Did the new theory not start off with two different gauge fields, which we indicate by the mathematical terms '$SU(2)$' and '$U(1)$'? That is not unification. On the other hand, however, you could emphasize that both force systems are based on exactly the same mathematical principles, and furthermore they are intimately mixed to produce the phenomena that are now explained. And, as in any meaningful theory, numerous new predictions emerged that could be vindicated by experiment. It was much less the case with the previous formalisms, so in this sense one could talk of unification.

Unification became an attractive buzz word. Physicists in previous decades had long been searching for the 'Unified Field Theory', the one theory explaining just about everything. 'Unification' had nearly been achieved, but not quite. Could we not do this a bit better?

If the three gauge systems – remember that we also have the strong force, based on $SU(3)$ – could be molded into one such system, this would imply a genuine improvement. Then you would not have three constants of Nature to describe these forces, but just one. Instead of twenty uncalculable constants, such a new theory would most likely have many fewer.

Life, unfortunately, is not this simple. The three constants describing the three gauge forces in the Standard Model are all quite different from each other. In particular, the constant belonging to the strong force, $SU(3)$, is many times bigger than that of the weak force ($SU(2)$), which in turn is bigger than the $U(1)$ constant. If all these systems were to be described in terms of one gauge field construct, all these numbers had to be more or less equal.

Now let us again make use of our imaginary microscope. Assuming that the Standard Model remains more or less valid at much smaller distance scales, we can calculate the strengths of the forces when particles approach each other at distances much smaller than hitherto considered. Perhaps you recall from Chapter 13 that the strong force becomes slightly less strong. This is also true, but to a much lesser extent, for the $SU(2)$ force. The $U(1)$ force can only increase in strength at smaller distances.

These changes in the relative strengths of the forces are only tiny. But let us now enter the super-highway in the desert with a fast car. We travel from the 1000 MeV entry to an exit at about 10 000 000 000 000 000 000 or 10^{19} MeV. This corresponds to a distance scale of approximately 10^{-30} centimeters. It was discovered, by Howard Georgi, Helen Quinn and Steven Weinberg, that this is the region where all three gauge coupling constants become equal. Is it a coincidence that all three become equal *simultaneously*? And is it a coincidence that this happens near the end of the highway? There are only three more zeros to go to reach a turning point that will be discussed in subsequent chapters.

Howard Georgi and Sheldon Glashow figured out how to write down a genuinely unified model in the energy domain of 10^{19} MeV such that, if you drive back along the highway, the three gauge forces as we know them re-emerge. Indeed they found just such a model. If we are willing to forget for a moment the fine-tuning problem, we have here a fantastic theory. The formula is '$SU(5)$'. It means that the smallest multiplet must have five members, but you can also have multiplets containing ten particles. In Figure 22, the multiplets for the fermions, *and now also their antiparticles*, have been arranged in a neat pattern. Each generation has multiplets that are indicated by the symbols 5, $\overline{5}$, 10 and $\overline{10}$. The multiplet containing ten particles is the arrangement obtained by asking how many ways two *different* members of a 5-plet may be combined together. The $\overline{10}$ is obtained in the same way, starting from the $\overline{5}$. These new objects are then identified with the known particles following certain mathematical rules. And now there will be new forces that may transmute a particle into another particle in the same multiplet. Besides the known gauge bosons W and Z of the weak force and the gluons of the strong force, this theory proposes the existence of a new gauge boson, the X boson, that brings about these transmutations.

Notice that right-handed neutrinos and left-handed antineutrinos are absent, but these could be added to the scheme. They would be isolated into three *singlets*, which implies that none of the gauge forces act on them.

Those gauge forces that we are already used to are the ones that cause transitions in such a way that they do not cross the broken lines. The X boson, predicted by Georgi and Glashow, is predicted to give transitions across the broken lines. Its mass must be somewhere in the 10^{19} MeV region. A peculiarity of the X boson is that it can turn quarks into leptons, and even into antiquarks. Figure 23 is a Feynman diagram showing what may happen when an X boson is exchanged.

A consequence of this model is that the proton is not absolutely stable; that it can decay into two or more lighter particles. Now we have already seen in Chapter 17 that, in the older Standard Model, protons can also decay, but only three at once, and that decay is so rare in the present day universe that it will probably never occur again.† Here we have a decay process for individual protons which will take place more frequently.

A good theory is characterized by the fact that it enables one to perform accurate calculations and make precise predictions. Indeed in this model one can calculate how long it takes for a proton on the average to decay. It is hard to do this very accurately because the time span is enormous: the prediction amounts to 10^{29} to 10^{30} years. The reason why transitions as sketched in Figure 23 take so long is not difficult to work out. The X boson is so heavy that enormous amounts of energy are required to produce it. Therefore the transition becomes extremely improbable. The probability for a decay at a given moment is inversely proportional to the *fourth* power of the X boson mass.

How can one test such a prediction experimentally? No one has the patience to wait 10^{30} years (our universe is 'merely' around 10^{10} years old). However, if you take 10^{30} protons together, you may expect that, on average, several of them decay every year. A moderately sized swimming pool easily accommodates 10^{32} protons in the form of water. Nowadays, measuring devices can be made that are sensitive enough to detect the decay of a single proton, although the experiment would have to be shielded as much as possible against cosmic rays (energetic

† It did probably occur in the very early stages of the Big Bang.

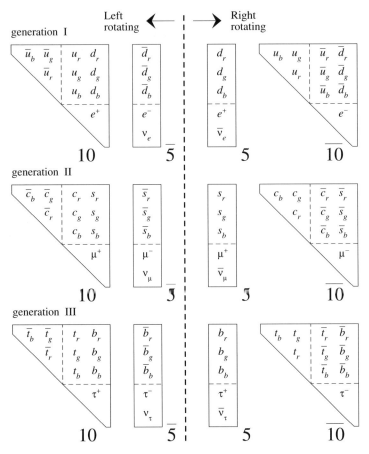

Figure 22. The multiplets in the Georgi–Glashow $SU(5)$ model. Right- and left-handed components occupy different positions. Bars on top of the particle names indicate the antiparticles. The quarks have three possible colors, indicated by the subscripts r (red), g (green) and b (blue).

pions and muons produced in our atmosphere by fast moving particles from outer space). Therefore one prefers to do such experiments several kilometers underground.

There is a limit to what one can measure. Cosmic rays also contain neutrinos, or produce them in our atmosphere. These cannot be stopped no matter how many kilometers of rock stand in the way. Neutrinos can cause interactions in water that are difficult to distinguish from a decaying proton. But by using great

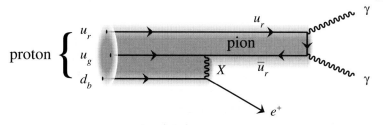

Figure 23. Proton decay by the exchange of an X boson. In this diagram this happens by the consecutive transitions $u_r\ u_g\ d_b\ \rightarrow\ u_r\ X\ \bar{u}_r\ d_b\ \rightarrow\ u_r\ \bar{u}_r\ e^+$. All fermions involved are in the first 10-plet of Figure 22. The X boson connects different parts of this multiplet. So here we see the decay $p\ \rightarrow\ \pi^0\ +\ e^+$. The pion then decays into two photons.

skill experimenters have been able to establish that the proton's average lifetime must be at least some 10^{31} years. This is the lower limit. No proton decay was identified with any certainty.

There are branches of science where such minor inconvenient discrepencies can easily be taken care of. You just stretch your theory a little. But here this does not work. The Georgi–Glashow $SU(5)$ model is out. It is possible to think of somewhat more complicated variations on the schemes that are not (yet) in contradiction with experimental data, but we get the feeling that we may be on the wrong track. For me, the main argument for this comes from the theory itself: the fine-tuning problem.

'Grand Unified Theories' of the $SU(5)$ type must contain at least two types of Higgs particles. The field of one of these is so strong that it is able to generate masses as big as 10^{20} MeV, the X boson mass. The other Higgs field is so weak that the W and Z bosons 'only' weigh 80 000 and 90 000 MeV, respectively. So one field is some 10^{15} times stronger than the other. Our problem is that in our theories the Higgs fields would tend to influence each other. How could it ever happen that one is 10^{15} times weaker than the other? Whence this enormous scale difference? Here also some people see a role for supersymmetry. If you try to 'protect' the large mass ratios by supersymmetry you get 'Super-Grand Unified Theories'.

23 Supergravity

The gravitational force is surely one of the most remarkable forces acting upon our elementary particles. You may remember that at the beginning of our journey towards the world of the small we observed that gravity is much less important for tiny creatures than it is for large ones. We used the example that, whereas a mouse can climb up a nearly straight wall, an elephant cannot. For atoms and molecules, and all the other particles we have discussed so far, gravity is practically a negligible phenomenon. But when we look at particles considerably smaller than the size of an atomic nucleus, we reach a turning point. Gravity acts upon the *mass* of the particles, whereas all other forces act on something we call 'charge'. The difference is that charge depends only very slightly on the degree of magnification of our microscope, whereas mass is connected to *energy*, and if we try to localize a particle in a smaller volume then, according to the rules of quantum mechanics, there will be more motion; the energy of motion (called 'kinetic energy') increases. This is why smaller distances correspond to higher energies, and hence also larger masses. When distances are so small that the movements become relativistic (i.e. close to the speed of light) the effects of the gravitational force gradually begin to increase relative to the other forces; however, they are still incredibly weak, and so they have a long way to go before thay can compete in strength.

Let us again enter the Great Desert (Figure 29), until we reach a region of particle physics where the energy (per particle) is much bigger than we can presently study in our laboratories. I should admit that the Great Desert is an area of speculation, and that we still know very little about it. We assume that the ground rules of particle physics remain valid. One thing, however, is abundantly clear: there will be an end to the road. One can calculate quite easily when it will happen that the gravitational force will overtake all the others, and this is where the Desert will end. I have already indicated where this boundary is: it occurs where the masses or energies grow beyond about 10^{19} proton masses, and this implies that we are looking at structures with sizes of about 10^{-33} centimeters.

We call this mass the *Planck mass* and we call the corresponding distance the *Planck distance*. The Planck mass can also be expressed in grams: 22 micrograms, which is the mass of a rather tiny grain of sugar (which thereby makes this amount the only Planck number that sounds more or less reasonable; the other numbers are totally outlandish!) This means that if you try to localize a particle with an accuracy of one Planck length, quantum fluctuations will give it so much energy that its mass will be as large as the Planck mass, and the effects of the gravitational forces among such particles will exceed those of any other kind of force. For these particles, gravity is a strong force.

And if gravity becomes a strong force, this is nothing short of a disaster. You cannot get away with the lamentation that this will make gravity there as difficult as 'quantum chromodynamics' is when it interacts with the quarks. Here the situation is much more serious. The smaller the structures are that you would try to study, the stronger this force would become, to the extent that even the roughest attempts to describe them will be jeopardized, and give completely nonsensical results.

Everything we now think we know about Nature will be invalid at the Planck scale. We thought we knew it all so precisely. Einstein's theory about the nature of the gravitational force works splendidly. It starts from a very fundamental principle, one that practically *has* to be right: gravity is a property of space and time themselves. Space and time are 'curved'. By 'curved' I mean exactly what happens to a smooth piece of paper after it gets wet: it becomes wrinkled, and there is no way to iron it flat. The gravitational force is to be attributed to such wrinkles in space and time.

The closer we home in on the Planck length, the stronger the need is felt to apply the laws of quantum mechanics to these space-time wrinkles. As long as the wrinkles are modest, this can be done, which results in a theory we call 'quantum gravity'. This theory predicts the existence of the aforementioned gravitons, elementary particles with spin 2 and zero mass.

The closer we get to the Planck length, the bumpier space-time becomes, simply because the tinier wrinkles are more pronounced than the larger ones. The usual *uncertainties*, typical for quantum mechanics, will make the wrinkles fuzzier. And if we try to go beyond the Planck length, everything goes wrong. The curvature, and the uncertainties in there, become so big that the notion 'distance between two points' no longer makes sense, because no measuring rods will fit in this space. Space and time themselves are becoming useless entities. The mathematical definition of what space and time *mean* hinges upon a definition of 'distance between points'. All of this will probably imply that all of our understanding of physics will have to be turned upside down before a useful description of this sub-Planckian world is found.

The last stop before such a thing happens is called 'supergravity', a complicated mathematical construction that manages to combine supersymmetry with the gravitational force. Again I cannot resist the challenge to explain in simple terms what I am talking about now. What is supergravity?

In Chapter 20, I mentioned the 'super-multiplets'. Particles with spin 0 or 1 are put in one multiplet together with spin $\frac{1}{2}$ objects. You then have 'supersymmetry'. Now one may ask for a theory in which a kind of gauge particles exist that cause transitions between particles and their super-partners; super-gauge photons, so to speak.

Such a theory indeed exists, but, as I have said, it is rather complicated. This is due to the fact that, if you change a particle into one with different spin, this really means you are affecting its space-time properties, not just 'internal' properties such as charge or strangeness. Quite a few physicists in many countries have worked on this idea, but it was so difficult that together they needed many years to obtain an airtight formulation. I am not doing them much justice by recapitulating their conclusions in just a few lines: not only a particle's spin changes if a super-gauge photon acts on it, but the particle is also displaced slightly. Displacing an object is also exactly what a 'gauge transformation' in the theory of gravity does. Other forces such as electromagnetism do not directly displace particles, they affect the wave functions of particles in such a way that the particles continue their way in a slightly different direction than before.

A consequence of all of this is that a super-gauge theory can only be formulated if it is combined with gravity. Hence we call the resulting formalism *supergravity*. The super-gauge particle is a particle with spin $\frac{3}{2}$, called the *gravitino*. It is the super-partner of the graviton, and we met it briefly in Chapter 20.

In '$N = 2$' supergravity we have two gravitino species; in '$N = 4$' we have four. There also exists an '$N = 8$' version, the most complicated supergravity theory, and by that perhaps also the most interesting one. In '$N = 8$' supergravity every super-multiplet should have all spin values from 0 to 2. To be precise, if you consider the rotating motion about an axis in a fixed direction in space, the spin can change from -2 to 2 in eight steps of one half unit. But there must be only one graviton (whose spin relative to a fixed axis is always 2 or -2), since in Einstein's theory only one type of gravitational force is allowed, and therefore we may only have one '$N = 8$' super-multiplet.

It is an attractive feature of this theory that no other particle types are allowed, and that this single super-multiplet contains so many particle species that all existing particles could be taken care of. So here again we have an example of a theory that allows no extension or addition of something else. Around 1980, enthusiasm for this construction became widespread. Suppose this is *the* theory, has physics then come to an end? Do we then have this all-embracing theory from which all other forces and particle types could be derived?

Let us go back another twenty years. In the early 1960s, Richard Feynman was among the first scientists to try to devise a consistent quantum theory of gravity. Feynman discovered the 'ghost particles', particles that do not really exist but that seem to emerge at intermediate stages of a calculation when one tries to find out what the effect is of multiple graviton exchanges. This pioneering work was continued by the Americans Bryce DeWitt and Stanley Mandelstam, and the Russians Ludwig Faddeev, Victor Popov, Efim Fradkin, Andrei Slavnov and many others. Feynman's ghost particles are now part of the standard equipment of all particle theories.

Once Veltman and I had learned precisely how to renormalize the Yang–Mills theory, we also joined in the discussion about gravity. What we discovered was that the first effects of multiple graviton exchanges (diagrams with one closed loop in them, see Figure 24(b)), can be computed unambiguously, if you only look at gravitons. But if they are allowed to interact with other particle types, or if the exchanged particles form two or more intertwined loops (Figure 24(c)), we hit upon the first difficulties: more and more infinite (counter) forces have to be introduced to obtain finite, and hence meaningful, final results from the calculations. But these counter forces themselves lead to even worse infinities when they are included in loops. Things grow out of our control. We call such a theory unrenormalizable.

So how about supergravity? Here, at first, things look much better. Even at the level of three loops nothing seemed to go wrong, and enthusiasts already exclaimed that this could be no coincidence, and that the ultimate theory of all forces was in sight.

A theory of all forces: can one imagine such a thing? Could an exact formulation of the laws of physics be possible, and is it conceivable that we physicists will ever find this out? Some of us think that such a 'Theory of Everything' is in sight. Other physicists say that even the thought of one is an extreme display of arrogance from a short-sighted minority. We will return to this theme later. Let me first try to pin down the considerations on which this apparently absurd 'arrogance' is based.

You could expect me to follow Gulliver further along his route towards the

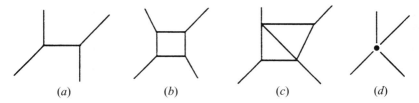

Figure 24. (*a*) Simplest particle exchange process. In this kind of diagram the effects of quantum mechanics are still minimal. (*b*) Multiple exchange resulting in a single closed loop. (*c*) Two loop diagram. In quantum gravity these exchanges would only give finite effects if a new, infinite interaction of type (*d*) is introduced. But if we insert that interaction in one of the vertices of (*b*) or (*c*), the complications run out of control.

world of the even smaller objects. Alas, I am afraid that travelling any further will be impossible, because in this realm *space and time cease to exist!* You cannot talk of two points closer to each other than the Planck length, because the curvature, the wrinkles, of the region in between can no longer be measured. Stephen Hawking once suggested that space and time at this scale are so very wrinkled that they form a kind of foam, with particles like protozoans swimming around in the soap films. To go from one place to another they have many routes to choose from. But even this picture is too simple: because it is impossible to determine distances or sizes, the little bubbles would be indistinguishable from the big ones.

The most radical viewpoint – one most physicists would be unwilling to adopt – is that space and time only consist of a collection of isolated points; particles can be on those points, but not in between. Actually this would be the most logical conclusion to draw, since by 'quantum fluctuations' all points where particles can be must be separated from each other by at least one unit of Planck distance. But we will not get away with this too easily, because how then do we explain that these points hang together to form the fabric we call space-time?

We actually have no idea how to answer such questions. But why should it be impossible to find a mathematically consistent formalism for all these mathematical aspects of space-time? It seems to me to be a challenge that might keep humanity busy for many more generations, but why should it be impossible to find the correct answer? Perhaps it is just an illusion, but it seems very much as if Nature is built from little Lego™ blocks, a construction kit with units the size of the Planck length with no continuum in between. Maybe the rules of the game do fit in a science text. And that then would be the Ultimate Universal

Theory I hinted at in the beginning of this book. If such a theory exists, we will find it, sooner or later. It is the Theory of All Forces, indeed the 'Theory of Everything', that many physicists are dreaming about, even though they are often led by very different intuitive arguments.

24 Eleven-dimensional space-time

We return to the next-to-last bus stop on the super-highway, called 'supergravity'. Supergravity theory worked well, but not well enough. At some points the mathematical construction did not work perfectly. Also, not all particle types seemed to fit together in such a model, and not all infinities cancelled. In courageous perseverence, researchers next tried the same theories in spaces with many more dimensions than ours.

A two-dimensional space can be compared with the surface of a piece of paper, such as the pages you are reading now. Suppose you tear one page out now and roll it up to obtain a cylinder. For a little red spider that happened to be walking on your paper, this makes little difference. It takes a long time for the spider to walk one full circle around the cylinder. It probably would not even notice that it came back to where it started from. We say that its world is still two-dimensional. But, seen from some distance, the pipe is rather like a stick, having only one dimension. In the same vein, the world of the very tiniest particles could have *more than three* (space-like) dimensions. These tiny particles could be like our little red spider. They would not notice that some of their dimensions are 'rolled up'. For us, these rolled-up dimensions have become invisible. This idea had already been suggested by Theodor Kaluza in 1919, and was further elaborated upon by Oskar Klein, in Stockholm, Sweden. And they discovered something else. The component of the gravitational field in the direction in which space is curled up obeys exactly the same laws as Maxwell's laws of electromagnetism! Could it be that electromagnetism is nothing but gravity in a rolled-up dimension? Einstein was enthusiastic when he heard about this idea, but it was soon realized that there is nothing one can predict with such a theory, and it was abandoned.

The supergravity experts rediscovered this idea of Kaluza and Klein. Once we start considering many extra dimensions, we enter into a Valhalla of mathematics where we can roll things up in many different ways. The components of the gravitational force fields in these rolled-up directions now act as various kinds of

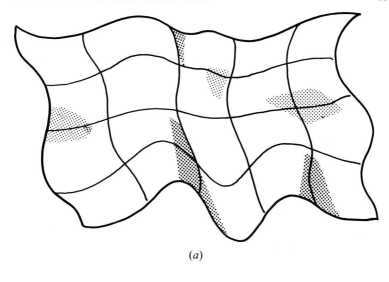

(a)

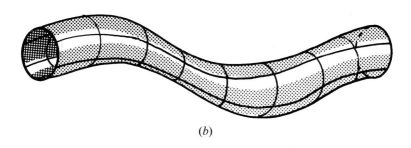

(b)

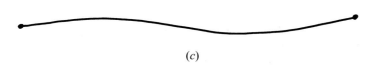

(c)

Figure 25. (a) Curved space. Two dimensions have been sketched. (b) Two-dimensional space, of which one dimension has been curled up. (c) One-dimensional space.

gauge fields. We obtain, practically for nothing, not only electromagnetism but also other gauge forces. The magic number became eleven dimensions. Three of these form ordinary space, and one is time. The seven remaining ones are rolled up. By a certain amount of trickery with the numbers, this system was found to have a bigger symmetry than good old four-dimensional space-time. The observed fields and particles could now be accommodated easily. The larger symmetry meant that unwanted infinities cancelled out against each other with more perfection than before.

True, this picture seemed to be quite the opposite of the notion that space and time are perhaps nothing more than isolated points, since then the entire notion of 'dimensions' does not make much sense any more. But mathematicians are not abashed by such apparent contradictions. According to them, there are all sorts of relationships between curled-up spaces and the mathematics of whole, 'loose' numbers (one could indicate isolated points of space-time by integers). Could it be that there exist several ways of describing our space and time, all mathematically equivalent? We simply do not know.

It is my suspicion that eleven-dimensional supergravity may, at its best, be just the tip of one marvellous iceberg, or else it could simply be wrong altogether. We should not forget at this stage that we are dealing here with wild ideas, and that the theoretical arguments for them are, as yet, still extremely weak. Why supersymmetry? Why eleven dimensions? Because this would make everything so beautifully symmetric? And above all, why still a continuum, if we already know that space and time lose their usual meaning at ultra-short distances? A persistent difficulty in this kind of theory would also be that the inter-particle forces are always treated as *perturbations*, affecting the particle trajectories which otherwise would be perfectly straight. But then there will be new (and different) perturbations upon these perturbed trajectories, and perturbations on those, and so on. This series of perturbations never ends, and this problem stands very much in the way of any attempt towards an exact formulation. True, this problem also affected the old 'Standard Model', but there at least one could argue that, where it really mattered, the forces could be kept small, so that the series of perturbing effects rapidly converged. This could no longer be so in our (super-)gravity theory; at small distances, the forces become strong.

I admit that I was relieved when the first serious difficulties in this theory showed up, and it turned out not to be possible to have infinities cancelling in diagrams with more than *seven* closed loops in them. The theory (or rather the speculation that this by itself would be the 'Theory of Everything') was abandoned, because something nicer appeared on the horizon.

25 Attaching the superstring

The superstring saga has its roots in the prehistory of particle physics: the 1960s. In Chapter 13, I recounted how Gabriele Veneziano toyed with formulae for the strongly interacting mesons. It took several years before it became clear that these are exactly the expressions obtained if each of these mesons is viewed as being a kind of rope with a quark at one end and an antiquark at the other. The ropes can be stretched *ad infinitum*, because stretching them adds energy to them, which will be turned into matter: that is, more rope.

The reason why Veneziano's formula described the properties of mesons so well was that this really is what mesons look like, approximately. Except that the ropes are not infinitely thin, they are fat ropes, formed by the pattern of strong force lines between the quarks. At higher energies Veneziano's formula becomes less accurate because features at smaller distance scales are being probed and then we see that the flux tubes produced by the strong force are no longer like strings. Rather than Veneziano's model, 'quantum chromodynamics', that is, the $SU(3)$ color gauge theory, was bestowed with the honor of being regarded as the prime theory for the mesons and the baryons.

But this did not imply that Veneziano's expressions were forgotten. Could one not construct an alternative theory for some kinds of particles that really consist of ideal, unbreakable 'strings'? In the 1970s physicists began to investigate whether the theory of mutually interacting strings could be improved.

In principle the philosophy was simple. Up to now all particles in any version of the Standard Model have been regarded as being point-like. If a quark or a lepton deviates from behaving like a point, it is merely because it becomes surrounded by a thin cloud of other point-like particles. Particles can interact in the first instance only if they sit exactly at the same point in space-time – indirect interaction takes place when two particles exchange a third particle such as a gauge photon. Also, and this is a related fact, force fields associated with such particles can be 'observed' or experienced at every single point in space-time separately.

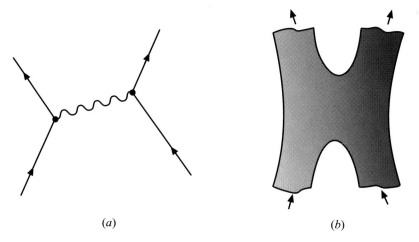

Figure 26. (*a*) Feynman diagram. (*b*) String diagram: the diagram indicates how the end points of string move about in space and time. In contrast to case (*a*), one cannot say where exactly the 'interactions' took place.

This, it was argued, we will do differently now. The next mathematical concept after the 'point' is the 'curve', or simply some arbitrary curved line moving about in space and time according to certain rules. Intuitively one could understand that interactions between point-like objects should be unnatural, because how could they ever find each other? Demanding that interactions only take place when two points exactly coincide should inevitably lead to infinities, as indeed happens in 'ordinary' field theories. It is much easier, however, for curves to meet each other somewhere and subsequently to undergo some sort of exchange process.

For the simplest string theories, this reasoning is not quite right; in these, interactions take place when two end points are joined together, or, conversely, when a string breaks. And, using the same arguments as before, this does not seem to be quite natural. Nevertheless, an improvement is obtained in comparison with point particle theories. Following an interaction process between strings in space-time, see Figure 26, we see how Feynman diagrams are replaced by 'string diagrams', of a more elegant appearance. Arguing that most of the difficulties in ordinary field theories arise from the fact that particles are forced to coalesce at special points in space-time, one might suspect that string theory will be free of such difficulties. In Figure 26(*b*), it may be seen that such special points, or vertices, are absent.

But string theory was not finished yet. Just as elementary particles are able to

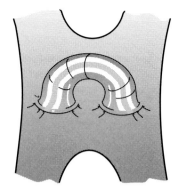

Figure 27. Higher order string diagrams.

produce 'loop diagrams', as in Figure 24, string diagrams should also be allowed to form more complicated patterns. During an exchange process, two strings could interact once more, and then diagrams are obtained such as in Figure 27. To calculate the effects of such diagrams was not at all an easy task, and the rules for such calculations had to be designed from scratch. Just as Richard Feynman had had to formulate the calculational rules for loop diagrams in gauge theories, it became necessary to repeat this process once again for string theory. The first results brought good news and bad news.

First the good news. Those nasty infinite expressions that forced us to produce lengthy arguments in formulating the former quantum field theories now disappeared: we were dealing exclusively with healthy 'finite' mathematical expressions in our formulae. Now, is this really good news? Had not we already learned how to deal with the apparently infinite outcomes of the older field theories? All we had to do was to be sufficiently careful not to talk about unobservable things such as 'naked charges' and 'naked masses', which were ill-defined anyway; but it had turned out to be perfectly possible to make accurate predictions that could be tested experimentally, such as collision probabilities. Well, apparently life has become a bit easier for the string theorists. And, as a bonus, we see that the theory stays manageable even if space and time have more than the usual four dimensions, à la Kaluza–Klein. In more than four dimensions none of the standard quantum field theories can deal with the resulting infinities, that is to say, none of them is renormalizable. The string theories can be conveniently combined with the cute little games of Kaluza and Klein.

But now the bad news. The calculational rules did not work entirely correctly. Just like the gauge theories, for which Feynman had discovered the 'ghost particles', string theory also turned out to possess ghost solutions. The only way to get rid of these seemed to be to choose the string parameters in a very special way. But then different kinds of solutions appeared that could travel faster than light. Well that would be just about as bad. Perhaps science fiction authors think they know how to handle particles moving faster than light – sitting on (or in) such objects one would need no time at all to travel comfortably to distant star systems – but for serious physicists such particles are disastrous. Less scrupulous authors did indeed allow themselves to become involved with calculations on hypothetical particles of this nature, and they called them *tachyons* (Greek ταχύς = fast). But, according to the laws of quantum mechanics for elementary particles, a theory with tachyons in it would imply that empty space (vacuum) would *not be stable*. Such a theory is useless.

So there was work to do for a small group of tenacious supporters of string theory. And they were there. The mathematics of this theory seemed to be too beautiful to be left untouched, and the challenge to improve the theory in such a way that the tachyons would disappear was too tempting to ignore. True, the theory did allow for solutions in which little chunks of string move about like tachyons, but there were only a few species, one with spin 0 and one with spin 1. And we did get something else in return: it turned out that there were other, massless, string configurations with spin 0, 1, *and one with spin 2!* These were not tachyons. The massless spin 1 particle turned out to behave exactly as a gauge photon, and, most remarkably, the spin 2 object behaves exactly as a graviton. Its interactions were mimicking exactly the gravitational force. From a physical point of view, the reasons for this are quite simple: the only symmetry structure of a theory with interacting massless spin 2 particles known is that of gravity theory, so it simply could not have been otherwise! But this did imply that string theory would *automatically* generate a gravitational force! String theory would not only account for all those particle types that we had already observed, but also for the gravitational force, and, apparently, our humpy space-time is inevitably an integral part of this theory.

And so it happened that string theory came to be known as a possible candidate for a theory that would solve all our difficulties with the gravitational force, indeed, the gravitational force in this theory has already been unified with all other forces. This is, however, a version of string theory that has nothing to do anymore with the version that Veneziano had had in mind for describing mesons and the strong force. These are strings that do not have the sizes of protons and pions; they must be as small as the Planck length, which is about eighteen decimal places smaller. The tension strength in these strings is not 14 tons as in the 'string' connecting quarks, but a fantastically much greater figure (some thirty-six extra zeros). Only in this way could the graviton-string reproduce a gravitational force that would be *sufficiently weak*!

I remember vividly a discussion with John Schwarz at the California Institute of Technology at Pasadena somewhere around 1978, in which John tried to convince me of the beautiful promises of string theory. This was at a time when most theoreticians were still busy with supergravity theories, but he was already dreaming about string theories that could be the T.O.E ('Theory of Everything'). 'Suppose now that we could get rid of those ghost solutions and the tachyons some way or other,' he said. 'Wouldn't you think that that could be the ultimate theory?'

I was skeptical about this, but I could not talk him out of it. Maybe that was just as well, because, in around 1984, enthusiastic reports came from the United States, which were soon confirmed by more and more investigators afflicted by a new epidemic of discoveries. John Schwarz, together with Michael Green from Queen Mary College, at the University of London, were the proud fathers of a new method to combat the ghosts. The answer was a very special choice of an internal symmetry structure, and all that in a twenty-six dimensional space. Twenty-two of these twenty-six dimensions had to be curled up as prescribed in the Kaluza–Klein theory that I explained in the previous chapter.

The mathematics was further worked out by a young ingenious mathematical physicist from Princeton University, Edward Witten, who, together with Schwarz and Green, wrote a heavy two-volume book about the subject. It was also discovered that there should be *supersymmetry* on the string, but then the twenty-six-dimensional world would have to be replaced by one with *ten* dimensions, of which, of course, six would be curled up. Supersymmetry originated from the fact that there were also fermions attached to this string, like beads on a necklace. In itself this idea had already been around for some time – after all, somehow one would have to explain the existence of fermions – but the discovery that all difficulties could be eliminated simultaneously in this ten-dimensional string was new.

At Princeton, David Gross and his collaborators discovered a theory in which all the fermions on a string were allowed to run only in one direction, provided that all undulatory movements of the string that run in the same direction live in a ten-dimensional world, and that the movements running in the other direction live in a twenty-six dimensional world. He called this the 'heterotic string', and what is nice about it is that the left–right asymmetry arising from this greatly resembles the left–right asymmetry we observe in the weak interactions.

Enthusiastic superstring supporters were talking of a new era in physics, and all but introduced a new calendar: 1984 was to be the new year zero. 'The best theory since the discovery of quantum mechanics', it was said. Indeed, superstring theory exhibited startling properties. First, there were no infinite expressions anywhere so that the renormalization procedure, still difficult to accept for many, had become superfluous. Secondly, the gravitational force had become an unavoidable and inseparable ingredient of this theory. Just like all other 'elementary' particles, the graviton could be viewed as a small closed loop of string material. And, like the strings on a violin, a superstring can be made to vibrate in all sorts of ways. Most of the vibrating (and rotating) modes of the string are very heavy; they obtain masses somewhere close to the Planck mass. But, partly due to the existence of quite a few 'curled-up' dimensions, there are a considerable number of solutions to the equations for the string that represent particles with very low mass. These then could be identified with the various existing particle types as listed in the Standard Model.

A third, very important, aspect of superstring theory was that one could not simply add any further particle types to the theory. It is a 'package deal': you either accept the entire theory, or you reject it. Other particle types could not interact with this string, not even gravitationally. It was predicted that all particle properties should be completely computable, because the theory contains no freely adjustable parameter at all. So here we have another example of what I earlier called 'holism': Einstein's relativity theories tolerated no exceptions to their principles, and neither did the rules of quantum mechanics.

Supersymmetry plays an important role in superstring theory. Many investigators expect, therefore, that if it is calculated how the Great Desert will look there will also be an important role for the former supergravity theory. But that has now been downgraded from a fundamental theory to a possibly useful description of an intermediate stage.

Perhaps now you understand why superstring theory was seen as the forerunner of a 'Theory of Everything'. No word could be added edgeways. Our journey to the world of the small ended here. Only a few technical details remained to be sorted out.

If this book were to be put on a screen, this would be the moment for an ominous background tune to turn into a tell-tale crescendo. These 'few technical details' turned out to be quite obstinate. First, there were the curled-up dimensions. Curling them up could be done in a large number of ways. If you have two dimensions to be rolled up, you can choose whether you roll them up separately into pipes, or you can take the two together to form a single football.

Now, in the heterotic string we have, in one direction, six, and in the other direction twenty-two, dimensions to be curled up, and the number of ways this can be done is gigantic. Which of these did Nature choose? 'We can compute everything', they said, 'so this we'll also figure out'.

But they didn't. Not everything could be computed, restrained as we are to the mathematical techniques known at present. The problem is that, just like all other particle theories, string theory is a *perturbation theory*. The interactions are treated as perturbations upon the movements of strings that otherwise would have followed straight trajectories. To find the exact movements it is necessary to perform an unending series of calculations, which not only become harder and harder, but that almost certainly will not converge to one single answer. This means that there will always be a stage at which the computation of one more perturbative correction actually leads to a worse result than before.

Strictly speaking, this means that we have no theory at all. We also had this problem in ordinary particle theory, but it was not such a disaster. In many cases, one can show that the stage at which the perturbation series no longer converges is so far away that the 'approximative' results are already extremely accurate. Indeed, the celebrated 'Standard Model' is not *infinitely accurate*! But for the Standard Model this was rather an academic problem. Practical difficulties in performing accurate calculations are usually many times more serious.

Unfortunately, for string theories, this problem is disastrous. Most things one would like to calculate cannot be obtained from the perturbation series at all. How are those extra dimensions curled up? What are the masses of the lightest particles? (These numbers are extremely small compared with the Planck mass of about 22 micrograms.) And so on. There was no lack of ideas in the camp of superstring theorists, but reliable answers could not be given. Many attempts were made to construct a 'string-field theory', which is intended to bypass these problems. How one can hope to succeed along such lines, I do not know. In ordinary particle theories we have been working with fields for a long time, and yet we cannot avoid the need to work with perturbation expansions.

Actually, I would not even be prepared to call string theory a 'theory' but rather a 'model', or not even that: just a hunch. After all, a theory should come together with instructions on how to deal with it to identify the things one wishes to describe, in our case the elementary particles, and one should, at least in principle, be able to formulate the rules for calculating the properties of these particles, and how to make new predictions for them. Imagine that I give you a chair while explaining that the legs are still missing, and that the seat, back and armrest will perhaps be delivered soon; whatever I did give you, can I still call it a chair?

But I should not be this unfriendly. Had it not been that a large number of extremely able and energetic young theoretical physicists focused all their attention on this promising approach, I would have considered working on it myself. The idea is beautiful and far-reaching, but one should not underestimate the amount of work that still has to be done to turn this into something really useful. All sorts of things could go wrong. Now that I do not have any share in this, I can disclose in all honesty what my suspicions are: (super)strings are (again) an intermediate stage. The true Theory of Everything will be based on entirely different principles, but, in the mean time, research on strings, superstrings and heterotic strings will yield valuable tools for our quest towards such a goal.

26 Into the black hole

If superstrings will not yield the ultimate answers, then in which direction should we continue our research? Or did we throw ourselves so far into the world of the unknown and unintelligible that we are about to drown in nonsense? Are we burying ourselves beneath so many impossible questions that we should be considered lost for science? Does it make any sense at all to speculate about a Theory of Everything in this strange world of Planck numbers? Perhaps the title of this chapter makes you fear the worst.

Nothing excites our curiosity more than the unintelligible. What is so curious about the world at the Planck length is that *no model at all* can be found that gives a reasonably self-consistent description of particles that influence each other with such strong gravitational forces, while at the same time obeying the laws of quantum mechanics. So, even if we had been able to perform experiments with particles that hit each other with Planckian energies, we would not have known how to compare the results with a theory. There is work here to do for physicists: make a theory. We do not care too much how such a theory describes the gravitational force, but there are quite a few demands on our list that make the creation of a candidate theory extremely difficult. As I explained at the end of the preceding chapter, superstring theory came close to doing just this, but it may well fail to fulfil its promises.

In the very first place, a theory must be mathematically accurate and enable us to perform accurate calculations about the behavior of particles under all imaginable circumstances. I often receive letters from amateur physicists who try to sell me the prettiest ideas which unfortunately are useless to us because their prescriptions do not qualify in rigor and precision to the kind we have become used to and which we now require naturally in all our theories.

Secondly, we naturally want the theory to handle the gravitational force in such a way that agreement is obtained with the way it is formulated in Einstein's general relativity theory. We know that the gravitational force between heavy bodies such as stars and planets obeys this theory very accurately (this has been confirmed dramatically in the observations of pulsars, fast rotating compact stars, by Russel Hulse and Joe Taylor during the last two decades). Our candidate theory should conform with this.

Thirdly, we know that the laws of quantum mechanics are inexorable. So we want our theory to be formulated in terms of the doctrines of quantum mechanics. Both quantum mechanics and relativity theory have the property that, as soon as one admits even the slightest deviation from their principles, a totally different theory would result that in no way resembles the world as (we think) we know it. 'A little bit relativistic' or 'a little bit quantum mechanical' are as meaningless as 'a little bit pregnant'. We could imagine, on the other hand, that quantum mechanics, or general relativity, or both, could be frameworks which are too restricted for the new theory, so that their principles should be further extended. But we cannot ignore them. I do not plan to annoy you with attempts to explain theories we do not even understand ourselves, but I can try to indicate which paths there may be that can be followed further.

In our universe, many stars exist whose masses are considerably larger than that of our own star, the Sun. Because of this, the gravitational force on their surfaces is considerably stronger than on Earth or on the Sun. The enormous quantities of matter in such a star cause an unimaginably high pressure inside it. It is only because the temperatures inside such a star are usually also gigantic, that there is sufficient counter pressure to keep the star from collapsing. The star, however, loses heat. At the beginning of a star's life, there will be all sorts of nuclear reactions that keep the temperature high and even cause it to increase, but, sooner or later, the nuclear fuel runs out. The heavier the star, the higher the pressure and temperature, and the quicker the fuel is burnt away. The counter pressure cedes and the star collapses under the pressure. As the star decreases in size, the gravitational force increases even further, and an implosion – a sudden and complete collapse – can no longer be avoided.

Often this implosion releases so much heat that the star's outer layers are blown away by the radiation pressure. Then the implosion can be brought to a halt, and an extremely compact sphere consisting of 'nuclear material' results: a so-called neutron star. These neutron stars often show an impressive rate of rotation (greater than 500 revolutions/second), and irregularities on the surface cause an equally rapidly pulsating radio signal. The observation of such a radio signal was how these objects were first discovered, and that is why they are called 'pulsars'. In the astronomical tables, they are indicated by the letters 'LGM'. This is a relic from the times when the possibility was considered that these signals came from an extraterrestial civilization: LGM = 'Little Green Men'.

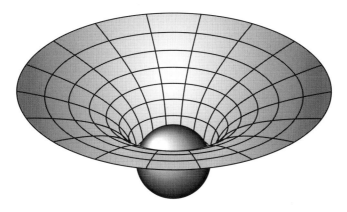

Figure 28. The black hole, an artist's impression.

If not enough material was blown away from the outer layers, or if new material lands on the pulsar, then even this solid sphere of nuclear material cannot withstand the gravitational pressure. Calculations first carried out by the astronomer Subrahmanyan Chandrasekhar show inexorably that if a compact, cool object is given a mass more than just a few times that of the Sun no species of matter, of any kind, exists that can withstand the pressure. The gravitational force becomes so strong that only Einstein's general theory of relativity can tell us what will happen. Since the gravitational force acts *collectively* on all particles in the star, it can still be considered as very weak where it acts on a single particle. So we do not (yet) have to fear that we would need *quantum* gravity to calculate precisely the next chain of events. Presumably John Archibald Wheeler was the first to realize what it is that *would* result from these events, and we have no doubts whatsoever that he was right.

The outcome of the events is what Wheeler called a 'black hole'. A black hole is what results if the imploding matter at a certain spot reaches the speed of light. A mathematical boundary line is passed, a point of no return. An (unfortunate) space traveller who enters the hole together with the imploding matter would, at that point, not be able to escape, even if he could turn around with the speed of light. With him, all signals he tried to emit would also be trapped, never to be seen again.

If one views all this from a safe distance then soon the signals emitted by the imploding material, and the astronaut flying in, would become too weak to be detected. The object turns black, so the name black hole is quite apt. The black hole will end up being nothing but a ball of 'pure gravity', and because of this its properties can be computed with mathematical precision. Just three parameters are sufficient to characterize the black hole completely: its *mass*, its *angular momentum* (amount of rotational movement), and its *electric charge*.

One can also calculate how beams of elementary particles behave when they venture close to a black hole. This is what was studied by Stephen Hawking in 1974. He hit upon a difficulty which he thought at first was a mere formality: when exactly do you call something a particle, and when can it be viewed as part of empty space? If something occupies a positive energy level, it is usually considered to be a particle, but when the energy level is below zero the level has to be occupied, and only the *absence* of an object in such a level is experienced as a particle, as we have already seen in Chapter 17. But energy levels are not uniquely defined when the surroundings are in motion, as is the case for particles in the vicinity of a black hole. Consequently, one may find that, whereas an astronaut falling into a black hole would think space-time around him was empty, an onlooker from the outside would seem to experience particles just escaping capture by the hole. Is this an inaccuracy in the theory? This is what Hawking first thought, but no matter how he tried to refine his calculations, he always seemed to obtain a weak flow of particles escaping from the hole. And at that point he made the discovery that I still consider to be his most important. These particles are real! Every black hole is emitting a constant flow of particles of all conceivable species.

The intensity of this particle emission is inversely proportional to the square of the black hole's mass. For those black holes that are discussed by astronomers, the so-called Hawking radiation will be so extremely weak that it will always be overshadowed by whatever falls *into* the hole, even if it is far away from other stars or star systems. But what counts now is the mere principle: something can escape from the black hole, and it escapes spontaneously.†

This now implies that a black hole can lose mass, and so it could, at least in principle, become lighter than the critical mass calculated by Chandrasekhar. And the lighter the black hole, the more efficient it is in losing even more mass. A quick calculation tells us that by the time a black hole has reached the mass of a fairly sizable mountain, it will only need a few more seconds to blast all of its mass away in the form of Hawking radiation. This would create a considerably more intensive explosion than the most powerful nuclear bomb.

---

† The particles emitted by a black hole follow the statistical rules of *thermal radiation*. The black hole has a definite temperature, and the radiation is the same as that of a snowflake, a lightbulb or a star. There definitely will not be little green men or astronauts among the emitted particles!

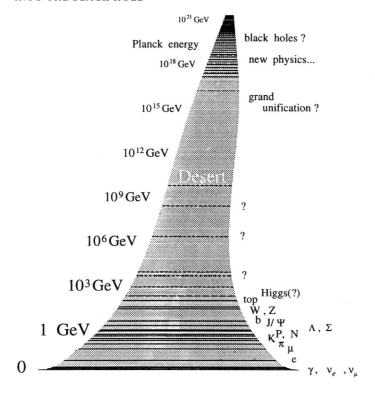

10^{21} GeV

Planck energy

black holes ?

10^{18} GeV

new physics...

10^{15} GeV

grand
unification ?

10^{12} GeV

Desert

10^{9} GeV   ?

10^{6} GeV   ?

10^{3} GeV   ?

top   Higgs(?)
W , Z
b   J/ Ψ Λ , Σ
1 GeV K P, N

π μ

0   e

γ , ν_e , ν_μ

Figure 29. The highway through the Great Desert.

True, astronomical black holes will never reach this stage, since they take in
more material than they emit, but in principle these processes are possible, and
we are interested in the details of what could happen. Perhaps I have suggested
that Hawking's calculation only made use of well-established laws of Nature,
and hence should be indisputable. This is not quite true, for two reasons.

First, we have never been able to study a black hole from close up, let alone
a black hole that is so tiny that its Hawking radiation can be detected. We do
not even know whether such mini black holes are present at all in our universe,
or whether perhaps they form an extremely rare minority among the heavenly
objects. Even if we think we know the theory, it would have done no harm if we
would have been able to check its predictions some way or another. The effect
Hawking discovered certainly is a quantum effect. Does everything then happen
exactly as we presently think it will?

And here I come to my second point. I could imagine a theory that gives another result. When I first became interested in this problem and began to produce a completely unambiguous description of the Hawking phenomenon, I tried to construct all sorts of alternative theories. I succeeded in formulating one that gives an outcome other than Hawking's: that the black hole would radiate considerably more intensely than he found. I admit that such deviating outcomes are only observed if the laws of quantum mechanics near black holes are handled differently from how they are considered when calculating conventional atomic phenomena, but I claim that there is no absolute proof that my deviating procedure 'is not allowed'.

By 1985, I knew Stephen Hawking from various conferences and workshops. I found the way in which he defies his serious physical handicap and manages to continue to play such an eminent role in present day theoretical physics unimaginable.† Today he can communicate with us very well thanks to an ingenious speaking computer that he punches with just one finger. With this finger he almost, but not quite, convinced me that my alternative theory is untenable. Now that I have studied black holes in much more detail I believe that his calculation is very probably correct. I just do not quite know yet for sure.

There is a much more important aspect connected to this Hawking radiation. The black hole decreases in size as it emits particles, and its radiation rapidly grows in intensity as the size reduces. Just before the final stages, the size of the black hole will be comparable to the Planck length, and the black hole mass will be only a little more than the Planck mass. The energies of the emitted particles will also correspond to the Planck mass. *Only a complete theory of quantum gravity will be able to predict and describe exactly what will happen to the black hole at that moment*!

This then is the importance of black holes for the theory of elementary particles at the Planck length. Black holes would be an ideal laboratory for thought experiments. All by themselves they reach the energy regime of the Planck numbers. A healthy theory should be able to tell us how to calculate here. And, for nearly a decade, I have been pointing out this objection to superstring theory: it tells us *exactly nothing* about black holes, let alone how a black hole that may have commenced its life as a large, 'astronomical' black hole, ends its life explosively. Since 1994, however, string theorists have been frantically trying to remedy this situation. As for any healthy theory of gravity, string theory does predict the existence of black holes, but, as of this writing, there is still much

---

† See Stephen Hawking, *A Brief History of Time* (New York, Bantam, 1988).

confusion concerning the procedures that are to be followed to calculate their physical properties.

And the more you think about it, the more important the role played by black holes in the world of the small seems to be. My starting point is presently: if you really want to understand how the gravitational force acts upon elementary particles, what better to choose as 'Gedanken' laboratory than the 'strongest conceivable gravitational field'? If one tries to localize any kind of particle with the precision of one Planck unit of length, the quantum uncertainty relation says that the uncertainty in its energy will be at least one Planck unit of energy. But then you generate mini black holes, whose sizes actually are larger than the Planck length. This is exactly the difficulty we have seen before: because of the humps, ripples and folds of space-time it becomes impossible to localize objects better than one unit of Planck length!

If the search for the ultimately small comes to an end, it is at the smallest possible object: a mini black hole. There, space and time lose their usual meaning, but what we should replace them with we do not know. Our search has ended, literally, as well as by way of speech, with a black hole.

27 Theories that do not yet exist...

And now I can report that implicating black holes in our considerations concerning the ultimate laws of physics means progress. I have now concentrated most of my own research towards this subject, asking myself this: suppose you have a tiny black hole and that you insist that it obey both the laws of quantum mechanics and those of gravity. Then how should one describe its behavior?

The shrewdness of this question is that I am assuming that an entire black hole should behave exactly as if it were an atom or molecule obeying the laws of quantum mechanics. Not everybody agrees with me here — not by any means. Some say that black holes are something altogether different. But what is so different about them? Black holes emit particles just like quite a few species of radioactive atoms do. Then why should they not do this by following the same rules? To put it more strongly, I happen to suspect that they absolutely *have to* obey such laws if you believe in any kind of 'law and order' at the Planck length.

One result of my calculations was a total surprise for me. I hit upon practically the same mathematical expressions as the ones in string theory! The formulae for capture and emission of particles by a black hole look exactly like Veneziano's formula. This is odd, since there is no question of strings. Now, as long as the theory is not finished, it is very hard to say whether and how string theory may perhaps be reconciled with black hole theory. In any case, both are incomplete; both may perhaps be the initial stages of something much more complete and beautiful.

There are quite a few researchers who have investigated their own ideas, which are quite different from mine. If you believe Stephen Hawking, black holes are just the beginning of much more serious deformations of space and time: for example, his idea of 'space-time foam', as I mentioned earlier. And that is not all. He and several others, in particular Sidney Coleman of Harvard University, speculate that a special role is played by 'wormholes'. A wormhole is a protrusion of space-time connecting two distantly separated regions of a universe, or it might even connect our universe with 'different universes'. Einstein's formulation of the theory of

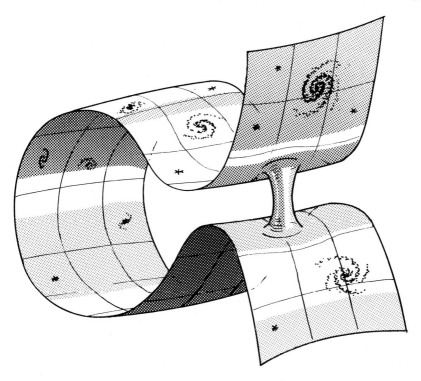

Figure 30. A wormhole may connect distant regions of a universe.

gravity could, in principle, admit such oddities. But, the investigators argue, if Einstein's theory allows wormholes, then they *have to* exist. This appears to agree with what we experienced while working with quantum mechanics: anything that is allowed is compulsory, that is to say, if some configuration is possible, it may actually occur with some definite probability. If you have read any science fiction, you will know that authors of that genre love wormholes. Imagine the endless possibilities: you jump into one, and *hop*, and in no time at all you pop up in Andromeda.

According to biologists, worms are useful since the holes they make are beneficial for the soil. Be that as it may, for quantum mechanics they are disastrous. Fortunately one can also interpret wormhole theory in such a way that wormholes are absolutely unobservable. I would rather keep it that way. Again one could suspect that wormholes, at best, are an intermediate phase of a better theory. Even if you refuse to take science fiction seriously, you might still imagine that, every now and then, a single elementary particle could slip through a wormhole from here to Andromeda. But the problem then is that the calculations made by Hawking and Coleman indicate that such phenomena are fundamentally *uncalculable*. Goodbye law and order. Curiously, neither Hawking nor Coleman draw from this the conclusion that wormholes must be forbidden. I do draw such a conclusion.

Yet another approach began with a formal analysis of quantum gravity by Abhay Ashtekar at the University of Syracuse in the State of New York. Expanding upon his ideas, the young investigators Lee Smolin and Carlo Rovelli proposed that the fundamental ingredients of space-time are not the points in there but rather the *closed loops*. This looks a bit like what we saw in string theory, but this is an altogether different approach; according to Smolin and Rovelli, the essential thing is that these loops are tied together in *knots*, and that outside the knots there is no space-time at all.

This is an attempt at constructing a new theory that I am following with interest – at last something that looks a bit like what I would want. In this theory, the only thing relevant is the number and the kinds of 'knots' linking the loops, and these are matters that in principle could be expressed simply by series of whole numbers. Just like loose points in space-time, except that they are naturally connected! Accidently, or perhaps not so accidently, *knot theory* is one of the most difficult branches of modern mathematics. During their excursions in string mathematics, the string supporters had also hit the knot problem. Edward Witten discovered during his analysis several new mathematical theorems about knots which earned him the prestigious Fields Medal.

If you think that these new ideas are no more than day-dreaming and a waste of time, I should disclose that what I have mentioned so far is not nearly the most extravagant attempts and approaches that have been published. In conferences and workshops that sometimes become like brainstorming parties, we are treated to far more bizarre concoctions. People philosophize about quantum jumps from one universe into another (usually through wormholes), about parallel worlds in 'quantum cosmology', about worlds in which the constants of Nature differ from ours, but which are connected to us by wormholes, and even about the question of whether you can create a universe inside a test tube. Such notions are beyond me.

28 Dominance of the rule of the smallest

Maybe the smallest structures in space-time should be pictured as 'superstrings', or perhaps they are pieces of cotton wool tied together as is advocated by Ashtekar and his followers. Maybe you tend to believe, as I do, that the dominant structures at the smallest possible scale are microscopic black holes. Each time, one conclusion seems to be inevitable: the *amount of information* one can store inside a tiny piece of space appears to be limited. Now, anyone who has worked with computers knows that information can be represented by series of zeros and ones. If an 'interaction' takes place, the zeros and ones are replaced by other zeros and ones.

Does this mean that the world we live in is nothing but a giant super-computer? Every book about the foundations of quantum mechanics will tell you that this would be too much of a simplification of matters. The laws of quantum mechanics, we read, are incompatible with any 'mechanical' explanation of what we see happening in Nature. Our future is not determined from the past by unambiguous, 'deterministic' rules.

This statement is based on a thought experiment invented by Einstein, Podolski and Rosen. It is an ingenious scheme designed such that quantum mechanics predicts an outcome that cannot possibly be reconciled with a deterministic theory. John Bell at CERN later turned this argument into an accurately formulated mathematical theorem. Thus we can imagine experiments for which the laws of quantum mechanics as we know them predict accurately what we will observe. It will be impossible to reproduce this prediction with any kind of deterministic theory, according to Bell. He did make one assumption, however: he assumed that information cannot spread faster than the speed of light.†

---

† Later, these kind of experiments were actually realized. As everyone expected, the quantum mechanical predictions were correct.

Here we have another example of a 'no-go theorem', a theorem that dictates with certainty how *not* to try to construct a theory, because it will not succeed.

As you may have noticed, I have too skeptical a mind to rely on no-go theorems. A similar theorem once told us that we should not try to combine internal symmetries among particles with their space-time symmetries, and that therefore it should be impossible to build super-multiplets of particles with different spins. So what about supersymmetry? Was that not a quite successful construction of super-multiplets containing different spins? Ah, well, that was the small-print of the theorem: it had been assumed that you could not put bosons together with fermions. One often forgets to mention the small-print, so that such theorems sometimes unjustifiably keep us from investigating important possibilities.

John Bell's theorems also contain small-print to which usually no one pays any attention. One of the possible escape routes, as I see them, is a theory in which what we call 'empty space' is not at all empty but shows a turmoil of activity. My colleagues will attack me on this. It is not at all easy to set up a good working 'mechanical theory accounting for quantum mechanical behavior', even if 'vacuum fluctuations' are taken into account. I admit that I also do not know how it should be done, but I suspect that there may be some escape route here.

I once prudently expressed these thoughts in a publication, and this brought me into contact with a group of people who, far away from the established views of mainstream physics, are convinced that Nature is an information processing machine. Their spokesman is Edward Fredkin. He is not a physicist, but a computer expert (and also an ingenious inventor). In computer technology it is common practice to work with models describing air movements in the atmosphere, or distributions of fish in the oceans, or anything else you can imagine, as if these were just ones and zeros. Such a computer model is often called a *cellular automaton*: numbers consisting of ones and zeros are stored in a large number of 'cells'. At the beat of a clock, the numbers in each cell are replaced by others following a precisely defined prescription, depending only on what the numbers in that cell were before, and on what the numbers in the nearest neighboring cells were.

Fredkin is convinced that our real world is also nothing but such a cellular automaton, albeit a gigantic one. One can follow the events in such cellular automata on a computer monitor. Beautifully colored patterns evolve before our eyes. By studying these patterns and their systematics, Fredkin had the idea that even quantum mechanical phenomena and forces resembling the gravitational force could be reproduced by these patterns.

But claims of this sort should be supported by hard evidence, and this is what Fredkin was unable to provide. He did make one curious and important observation: *it does not matter very much what exactly the program for the automaton is!* Since, he says, as soon as your program is sufficiently 'versatile' you can use any automaton to 'mimick' any other. You could, as it were, construct little computers with an automaton with which you can simulate the other ones. So he speaks of 'the universal cellular automaton'. I am one of the few theoretical physicists who do not simply dismiss such a representation of Nature as a huge cellular automaton by claiming it to be at odds with quantum mechanics. If you try to follow the patterns generated by the computer you see them quickly turn tremendously chaotic. If any apparent systematical behavior at all were to re-emerge at much larger scales (large compared with the sizes of the cells), then probably only the laws of statistics could be applicable here. And perhaps these are what what is presently called 'quantum mechanics'.

But then, claiming that the 'universal cellular automaton' should be the 'Theory of Everything', I fear, would be like throwing the baby out with the bath-water. How would one figure out what the properties of the elementary particles are? Half in the dark, we have been following what we thought to be a path towards the very smallest, and while groping our way we feel that our path continues no further. We lost our way back, and the car was parked somewhere along the highway through the Great Desert.

Nature as an information processing machine: is this an illusion? Is perhaps the mere idea not more than a natural product of our present culture? Is, after all, 'information' not one of the central pillars of modern society? Perhaps we are just being myopic. The future will tell. *If* the ultimate law of physics is one that only processes zeros and ones, then sooner or later humanity will figure this out; this much confidence in human ingenuity I do have.

Just imagine this to be the case, that there will be, one day, a rock-solid 'Theory of Everything', a fundamental law, the ultimate formulation of which is so simple and universal that no small changes, no amendments, are possible, a 'holistic' law – as we saw before. All properties of matter, all phenomena in space-time, all other laws of physics should be derivable from this one universal law. What effect would this have on physics, and eventually also on society?

The effect would be very small indeed. All but one of the branches of physics are *not* about discovering the T.O.E. To describe atoms, molecules, materials, gases, liquids and all other states of matter requires different mathematical techniques each time. Physicists investigating these topics are very well aware of the fact that the building blocks of these substances obey laws that are, in principle, known. What they are interested in is the question of how these laws conspire to produce the observed phenomena. This is the kind of question the T.O.E. is not capable of answering at all.† Therefore all these branches of physics will stay exactly as they are. Even for elementary particles, at all sorts of levels one will continue to ask the same questions as before. Or in other words: the end point of our route towards the smallest may perhaps become known, but we will continue asking questions such as: how does this route go exactly between *A* and *B*? Answering these questions will remain as difficult as ever, and huge laboratories for ingenious experiments will remain necessary and useful. Therefore the effects of a T.O.E. on physics will be quite remote.

As for the effects on society, these will be remote also, in the first instance. I am entertaining some hopes, or rather some naive illusions, that humanity will begin to understand a little bit better its place in this universe, and in particular that there is no place at all for all sorts of metaphysical mysteries.

Finding the so-called T.O.E. would be an unparalleled success of physics, and one might hope that such a success would stimulate the use of our successful procedures in some other sciences. I say this with some hesitation because one sees so many ill-conceived examples of this. If psychologists or sociologists begin to talk of *energy* or *entropy* they might unjustifiably pretend that in their fields such notions obey laws that are comparable to the ones in physics, or, even worse, that these would be defined with anything like the accuracy we have become used to in physics.

Probably more realistic was Richard Feynman's response when he was asked what to expect from a 'Theory of Everything'. Feynman said that he did not believe that it would ever come. 'But', he said, 'if it comes I would expect that something will happen not unlike what happens with a mountain top after it has been conquered by courageous and professional mountaineers.' These mountaineers may well be the last to enjoy the immense beauty of the unspoilt nature there. They will have been the first to make the top accessible. Easier roads will be made, and after that a cable car, and then the tourists will arrive. In the comfort of the newly built luxury restaurant on the top, these new 'explorers'

---

† Hence, the epitheton 'Theory of Everything' is quite deceptive. Indeed, I do not wish to be held responsible for it; the name was first heard from supergravity and superstring supporters who got carried away by their enthusiasm, and was subsequently picked up, of course, by the popular press to use it against them, whenever the moment seems to be there.

will air *their* vision of the mountain. The mountainneer will hardly recognize his discovery because of all the garbage on it.

Perhaps no one will ever have a complete view of the entire route from here to the very tiniest structures in the universe. Perhaps many more centuries will pass before humanity has completed the chart. Maybe the road exists, but we will never discover it, because of our limited intelligence. As stated earlier, this I find unlikely.

Could the road towards the tiniest structures be infinitely long? Could it be that there exists no 'tiniest structures'? It has been thought several times in history that the end of physics was in sight, and since this has always turned out to be wrong, it has come to be commonly accepted that a fundamental, universal theory will never be possible. Just to think that such a 'stone of wisdom' could be within reach is often regarded as a futility.

Let us first of all remove this 'stone of wisdom' aura. A universal theory would not at all imply that all natural phenomena would be explained. All we would know is a rather formal – though exact – series of equations which all phenomena would obey. The questions concerning the explanantions of these phenomena would now have turned into mathematical, calculation-technical questions. This is what most of those questions are already anyway, and they will not become in any way simpler.

Also, I hope to remove the impression that simple, exact laws of Nature would make our world so 'small' that there would be no place left for 'emotion', 'free will', 'life' or such things as awe and respect for the immense varieties of structures in our universe. The size of our entire universe must be expressed in terms of some 10^{54} Planck lengths. Its volume is that number raised to the third power (10^{162}). Its age is also something like 10^{54} Planck time steps. Such numbers contain a 'mere' fifty-four digits and hence, at first sight, seem to be small, but in reality these numbers are so tremendously large that there is more than enough place for all of the most miraculous features known and unknown to Man.

The only true resistance against a 'Theory of Everything' would be of a religious nature. Let us then for once read history† in the following way: humanity has been making discoveries one after the other that had been taken to be impossible before. Diseases, for instance, were once believed to be the instruments of the Gods until it was discovered that something could be done about them. During the ages of the great voyages of discovery, people continued to dream of oceans, continents and new civilizations still to be discovered, and

---

† Always remember: history will repeat itself, surely, but never in a predictable manner.

it seemed to be inconceivable that the time would come that every square meter of our globe would have been mapped out. Then the movements of the celestial bodies were thought to be the domain of the Gods (or that one God) until it was discovered that they could be understood, and that indeed these bodies are made out of the same materials as our Earth itself. Similarly, this universal law for all natural phenomena is presently in divine hands. For how long? Who can tell?

Glossary

angstrom   a distance unit equal to 10^{-10} meters.

angular momentum   amount of rotational motion of a particle or group of particles. To be measured in integer multiples, or an integer multiple plus one-half, of the unit $h/2\pi$, where h is Planck's constant.

antiparticle   particle type that is complementary to some given particle species. An antiparticle has an identical mass and spin to those of the corresponding particle, but electric charge and other quantum numbers, such as S, L and B, are opposite.

atom   building block of the chemical elements, consisting of a *nucleus*, which is built from protons and neutrons, with *electrons* in orbit around it.

B *see* baryon number.

baryon   collective name for particles that are sensitive to the strong force and have an amount of spin that is an integer plus one-half. Consists of three quarks.

baryon number (B)   number of baryons minus the number of antibaryons. Also the number of quarks minus the number of antiquarks divided by three.

Bose condensation   occurs when a large quantity of bosons attract each other and in so doing reach a total energy that is lower than without their presence.

boson   collective name for all particles with integer spin.

bottom   sometimes called 'beauty', the quantum number that specifies how many 'bottom quarks' are present.

C *see* charge conjugation.

CERN   European center for subnuclear research near Geneva, Switzerland.

charge conjugation (C)   replacement (for comparison) of a particle by its antiparticle (which has opposite electric charge).

charm   quantum number that specifies how many 'charm quarks' are present in an object.

color force   force by which quarks are attracted to one another.

conservation law   law of physics saying that the grand total of some quantity, such as energy or charge, will remain unchanged during any kind of interaction process.

coordinates   numbers that can localize the position of a point in a space, or the size and the direction of a vector.

Copenhagen interpretation   view according to which quantum mechanics yields all possible information concerning probability distributions. Descriptions in terms of particles and waves are 'complementary'.

Coulomb force   electrostatic force between two electrical charges. Proportional to the product of the charges and inversely proportional to the square of their distance.

coupling constant   constant of Nature describing the strength of a force or interaction.

decuplet   grouplet of ten members.

Dirac sea   infinite reservoir of negative energy particles populating empty space, according to Dirac's original theory. These particles are themselves invisible, but if one of them is *absent* this is observed as an antiparticle.

eightfold way   use of a mathematical method in quark theory that shows how mesons and baryons can be arranged in octets and decuplets.

electric charge   property of a particle that renders it sensitive to the electromagnetic force.

electric field   force field caused by the presence of electric charges. Affects the motion of particles with electric charge.

electromagnetic waves   pattern of electric and magnetic field lines describing many kinds of radiation, in accordance with the laws formulated by Maxwell.

electron   light, electrically negatively charged particle, which is abundant in all atoms. The electron is the lightest charged lepton.

energy   (for a particle with rest mass M and velocity v) this consists of a 'kinetic' part (energy of motion) approximately equal to $\frac{1}{2}Mv^2$, a part due to the rest mass equal to Mc^2 and a potential energy due to the presence of fields.

èta (η)   meson consisting of a quark and an antiquark, with spin 0. Fairly stable, even though the strong force can make it decay.

Fermi   unit of length equal to 10^{-15} meters.

fermion   collective name for all particles whose spin is half-integer, that is, $\frac{1}{2}, 1\frac{1}{2}, 2\frac{1}{2}, \ldots$ in natural units.

Feynman rules   scheme for calculations where diagrams are used to indicate the trajectories of particles.

field   quantity depending on where in space and time it is measured, such as air pressure, temperature distribution, etc. Also, the height of water can be regarded as a field. Oscillations in a field are called 'waves'.

field equations   laws determining the time dependence of a field.

gauge field   field of a 'gauge boson', a spin 1 particle transmitting a force comparable with the electromagnetic force, in which gauge invariance plays an important role.

gauge invariance   absence of any change when a gauge transformation of any kind is performed.

gauge transformation   change in our description of some situation that does not affect the physically observed phenomena we are describing.

generation of elementary particles   in the Standard Model each elementary fermion belongs to one of three generations. These generations are exact copies of each other except that the masses in the generations are different.

GIM mechanism   idea first proposed by Glashow, Iliopoulos and Majani suggesting that a fourth quark, 'charm', existed in order to explain the absence of weak interaction processes invoving the production of a Z^0 in low energy hadrons.

gluon   energy quantum of the strong force; a particle with spin 1 that transmits the strong force between quarks. Has eight possible colored states.

gravitino   gauge particle of supergravity theory. Has spin $\frac{3}{2}$.

graviton   energy quantum of gravity waves. Transmits the gravitational force. Has spin 2.

gravity   force with which heavy objects attract each other. Proportional to the product of the masses of the objects and inversely proportional to the square of the distance between the objects.

hadron   collective name for all particles sensitive to the strong force. Includes the baryons and the mesons.

Higgs mechanism (Higgs–Kibble mechanism)   asymmetric structure of empty space such that consequently all, or nearly all, particles obtain mass. See Chapter 11.

Higgs particle   boson with spin 0, not easily detectable yet populating empty space in huge quantities, giving rise to the Higgs mechanism.

interference   phenomenon occurring in waves and vibrations, in particular in quantum mechanics. If some process can take place along different routes, the various possibilities can either reinforce each other (positive interference) or attenuate each other (negative interference).

isospin (I_3)   conserved quantum number in collision events.

jet   little cloud consisting of several pions and nucleons moving in the same direction, often released in very high energy collision events. A high energy quark is always transformed into a jet.

J/Ψ   first hadronic particle to be discovered (in 1974) containing the charmed quark c, together with its antiparticle $\bar{c}$.

Kaluza–Klein theory   theory describing new dimensions of space and time that are tightly curled up.

kaon   lightest meson containing a strange quark s or antiquark $\bar{s}$. May be electrically charged or neutral.

knot theory   mathematical theory of knots and links. A closed piece of string can carry *knots* of different sorts; several closed strings can be tied together in *links*.

A knot or link is said to be different from another knot or link if one cannot be deformed into the other by moving the strings around or by stretching them.

L   *see* lepton number.

lepton   collective name for particles that are insensitive to the strong force and have spin $\frac{1}{2}$.

lepton number (*L*)   number of leptons minus the number of antileptons.

magnetic dipole moment   extent to which an electrically charged particle acts as a magnet. Light particles such as the electron are much stronger magnets than heavier ones such as the proton.

magnetic field   force field caused by fast moving electric charges, which affects, in turn, the orbits of other fast moving electrically charged particles.

magnetic monopole   particle carrying a magnetic 'north' charge not connected to a 'south' charge, or vice versa.

mass   extent to which a particle persists in its motion (inertia). Expressed in grams (g), kilograms (kg) or MeV/c^2 (1 MeV/c^2 = 1.78268×10^{-27} grams).

Maxwell theory   description of the equations obeyed by electric and magnetic fields

meson   particle with integer spin sensitive to the strong force. Built of a quark and an antiquark.

micron   distance unit equal to 10^{-6} meters.

mirror symmetry (*P*)   extent to which a physical phenomenon or a particle resembles its mirror image.

molecule   smallest unit of a chemical substance. Consists of several atoms.

momentum   mass multiplied by velocity.

multiplet   grouplet of particle types with comparable properties. For instance, the proton and the neutron form a doublet, and the three pions form a triplet.

muon   negatively or positively charged particle with spin $\frac{1}{2}$, about 200 times as heavy as an electron. Belongs with the electron in the lepton family.

naked particle   particle without the cloud of other particles that usually surround it. Idealization of a point particle.

neutrino   extremely light, possibly massless, electrically neutral particle with spin $\frac{1}{2}$. Since it is insensitive to the strong force, it is very inert. Belongs to the lepton family.

nucleon   collective name for proton, neutron and some of the heavier excited states of these particles (resonances).

nucleus   central part of an atom, built of protons and neutrons.

octet   grouplet of eight members.

parity (*P*)   *see* mirror symmetry.

Pauli's exclusion principle   this says that no two fermions can exist at the same point, or more generally occupy the same quantum state, unless they are in some other sense different from each other (different spin direction or different color, for instance). Fermions have to stay away from each other.

PC symmetry   resemblance between a particle and the mirror image (*P*) of its antiparticle (*C*).

phonon   energy quantum of sound. Relevant in materials at low temperature.

photon   energy quantum of electromagnetic radiation (light) with spin 1.

pion   particle with mass between that of the electron and the proton. Transmits the strong force between hadrons. Has spin 0. Can be electrically charged or neutral.

Planck scale   units for length (1.6×10^{-33} centimeters), mass (22 micrograms) and time (5.4×10^{-44} seconds) such that when expressed in these units the velocity of light, c, and Newton's constant of gravity, G, are equal to unity, and Planck's constant, h, is equal to 2π.

Planck's constant   constant of Nature relating the energy quantum, E, to the frequency, v, of any physical system: $E = h \times v$. Its value is
$h = 6.626075 \times 10^{-34}$ joule seconds.

positron   antiparticle of the electron.

probability   number, between zero and one, indicating how likely something is to happen.

proton   positively electrically charged building block of the atomic nucleus. Lightest baryon, and therefore the only stable one.

quantum chromodynamics   theory describing the strong 'color' forces between quarks in a hadron, based on the mathematical 'group' called $SU(3)$.

quantum electrodynamics   theory describing the electromagnetic interactions between electrons and photons.

quantum mechanics   theory describing the way small and light particles move, in which probabilities play an essential role. Energy, spin and several other properties of the particles are 'quantized'.

quark   building block of the hadrons. A quark is a fermion with spin $\frac{1}{2}$. Occurs only in multiples of three or is bound to an antiquark. Six species are known: u, d, s, c, b and t.

relativity, general   detailed geometric treatment of curved space and time, enabling one to understand the gravitational force, as discovered by Einstein in 1915.

relativity, special   geometric theory of space and time set up in such a way that the velocity of light appears to be the same for all observers, as discovered by Einstein in 1905.

renormalization   procedure amounting to a redefinition of the constants of Nature in such a way that the actually measured numbers are correctly reproduced. Often this implies that the original numbers that are to be put in a theory become infinite.

resonance   very unstable particle. If such a particle is temporarily produced in some interaction, it gives a peak in the number of collisions at a certain energy that looks like a resonance curve.

rest mass   mass of a particle when at rest. If a particle moves with a speed close to that of light, its mass appears to increase.

rho particle (ρ)   unstable meson with spin 1 and mass 770 MeV. May be electrically charged or neutral.

S *see* strangeness.

S-matrix   list of numbers, or formula, that describes the probability distributions of particles after a collision ('scattering') event.

sigma (σ) model   originally proposed as a model describing the interactions between pions and nucleons, where an extra particle called σ was introduced. Later also used to describe other systems.

spin   amount of rotational movement in a particle. More precisely, angular momentum. When expressed in multiples of $h/2\pi$ (h is Planck's constant) it is either integer (0, 1, 2, ...) or half-integer ($\frac{1}{2}$, $\frac{3}{2}$, $\frac{5}{2}$, ...).

strangeness (S)   quantum number for hadrons, defined in such a way that when added together the total strangeness of all particles before and after a collision will stay the same. It is also the number of 'strange' antiquarks $\bar{s}$ minus the number of strange quarks s.

string theory   depicts particles as little pieces of string – either open ended or in closed loops. Different particle types correspond to the different modes of vibration and rotation of a string.

strong force   force acting between the hadrons, which gives them their shape.

tau (τ)   new lepton. Comparable to the electron and the muon but much heavier.

TOE   'Theory of Everything'. Set of equations, including a manual for their use, that describes space-time, matter and the forces controlling all motion, with, in principle, infinite accuracy. Although some physicists, including the author, believe that such equations can be found, we do not seriously believe that they can be solved in any reasonable amount of time, so that the name 'Theory of Everything' for these equations is actually a euphemism.

top   also called 'truth', this is the sixth and heaviest known quark. Counterpart of bottom, recently also discovered experimentally at Fermilab near Chicago.

vacuum   empty space.

vacuum polarization   change of the properties of empty space caused by the strong fields of a nearby particle, because they have an effect on the particles and antiparticles in the Dirac sea.

Van der Waals force   universal attractive force between atoms and molecules caused by quantum mechanical fluctuations of the electric fields between them.

vector   physical quantity having a size and a direction in space. To be represented as a series of numbers (coordinates).

vector potential field   auxiliary field, first introduced in the theory of
    electrodynamics to describe electric and magnetic fields. Its components, one of
    which is the electric voltage, are only indirectly observable.

W particle   electrically charged carrier of the weak force, also called 'intermediate
    vector boson'.
weak force   force acting between all known particles. Very weak and with a very
    small range.
WIMPs   weakly interacting massive particles, an as yet unknown species of
    particles that may be present in huge quantities between the galaxies, producing
    gravitational fields there.

Yang–Mills field   *see* gauge field.
Yang–Mills theory   generalization of Maxwell's theory of electromagnetism. At
    least three types of electric and magnetic fields are present, and there is a gauge
    principle.

Z^0 particle   electrically neutral carrier of the weak force. Causes all weak 'neutral
    current' interactions.

Index